T0138945

Efficient Organic Light-Emitting Diodes (OLEDs)

Efficient Organic Light-Emitting Diodes (OLEDs)

Yi-Lu Chang

Pan Stanford Publishing

Published by

Pan Stanford Publishing Pte. Ltd.
Penthouse Level, Suntec Tower 3
8 Temasek Boulevard
Singapore 038988

Email: editorial@panstanford.com
Web: www.panstanford.com

British Library Cataloguing-in-Publication Data
A catalogue record for this book is available from the British Library.

Efficient Organic Light-Emitting Diodes (OLEDs)

ISBN 978-981-4613-80-4 (Hardcover)
ISBN 978-981-4613-81-1 (eBook)

Printed in the USA

Contents

Preface

Organic light-emitting diodes (OLEDs) are already prevalent in our daily lives in the form of mobile phone and tablet displays. More exotic products such as curved and wearable displays and even large-area TV panels have emerged of late. At the same time, intensive work is being carried out globally on OLEDs for design-friendly and energy-efficient artificial lighting. Moving toward this direction, we are expected to experience in our society a dramatic change of the magnitude no less than how halogen bulbs and fluorescent tubes had revolutionized our world several decades ago.

This book aims to provide a comprehensive understanding of this fast-growing subject. It is ideal for college students from multiple natural and applied science disciplines as well as industry professionals, especially from closely related fields such as inorganic light-emitting diodes (LEDs) and liquid crystal displays (LCDs). This book thoroughly explains the fundamental principles of key OLED concepts, buttressing them with simple mathematic formulations where necessary. It not only covers the most industrially applicable concepts such as top-emission OLEDs, white OLEDs, and tandem OLEDs, but also includes newer, advanced topics such as OLEDs based on thermally activated delayed fluorescence (TADF) and exciplex-forming co-hosts.

The book stems from my five years of research experience at the University of Toronto and as vice president of research at OTI Lumionics. In particular, thinking back to when I was first exposed to the field of OLEDs, I realized that there still wasn't a suitable book on the market that covered the subject in sufficient depth and, at the same time, explained it at the level convenient for college students and industry professionals in related fields.

I hope the readers will find this book fruitful and intriguing in the way it tackles complicated concepts, approaching them in a manner that is as simple as it is intuitive. Any feedback, inquiries, or reports of misprints from the readers of this first edition are greatly welcomed.

Yi-Lu Chang

April 25, 2015

Chapter 1

Introduction

Organic light-emitting diodes (OLEDs) are emerging as one of the most dominant display technologies and potentially as the next generation solid-state light source. This introductory book provides an up-to-date key research results as well as current industry standards of efficient OLEDs with thorough explanations on the underlying working principles. A number of the techniques presented are already being employed in commercial display and lighting products today. The focus of the book will be on the efficiency improvement strategies at practical luminance levels, which are more industrially applicable. This book is suitable for the general audience as well as for scholars and industrial professionals who wish to learn the state-of-the-art concepts and continue to develop the technology further.

1.1 Brief Overview of OLEDs

The organic light-emitting diode (OLED) has evolved rapidly over the past two decades to become the ultimate technology for displays primarily due to its unique, flexible, and thin form factor, as well as its ability to produce vibrant colors efficiently. Currently, active matrix organic light-emitting diode (AMOLED) displays are already prevalent in smart phones worldwide and are emerging in large-sized (77″) 4K ultrahigh-resolution televisions. As a display technology, it offers superior response time, low power consumption, and nearly

Efficient Organic Light-Emitting Diodes (OLEDs)
Yi-Lu Chang

ISBN 978-981-4613-80-4 (Hardcover), 978-981-4613-81-1 (eBook)
www.panstanford.com

infinite contrast since each pixel can be switched off completely. At the same time, it is also aggressively being studied globally for its potential use as the most desirable broadband, surface illumination source for general lighting. As a light source, it offers a large area, diffusive illumination based on environment-friendly materials, and enormous energy savings compared to traditional halogen bulbs and fluorescent tubes. Indeed, these fascinating features are projected to lead to a new world of artificial lighting. With recent breakthroughs in terms of material development and device architecture, there has never been a brighter future in OLED technology.

The first milestone in OLED technology that triggered a widespread community interest was reported by researchers from Eastman-Kodak in 1987.[1] It was a double-layer heterojunction organic electroluminescent (EL) device, shown in Fig. 1.1, that has a room temperature operating voltage of less than 10 V, a high brightness of over 1000 cd/m^2 (suitable for a commercial flat-panel TV display), and an efficiency of about 1%. Since then, studies were conducted globally in terms of both device structure and organic materials in response to the need for saturated red, green and blue emission colors from the display industry with higher efficiencies and longer lifetimes.

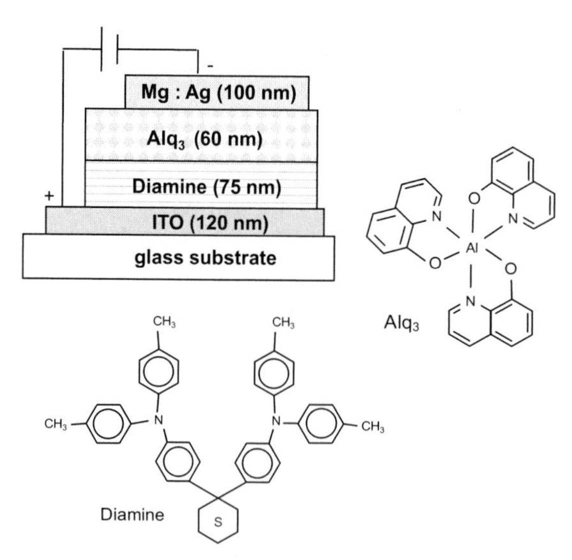

Figure 1.1 The first low-voltage OLED reported by Eastman-Kodak in 1987.

Another significant milestone in OLED device technology involves the discovery of phosphorescent emitters in OLEDs, which was first reported by S. Forrest and M. Baldo in 1998.[2,3] These phosphorescent emitters provide a significant boost in device efficiency and, as a result, have gradually become an indispensable emissive material for the flat-panel and portable display industry. Very recently in 2012, C. Adachi's group developed new emitters based on thermally activated delayed fluorescence (TADF)[3] to significantly lower the cost of the device, while maintaining efficiencies as high as those achieved from phosphorescent emitter-based OLEDs. More interestingly, the same concept of TADF has been further exploited as the basis for building efficient host layers using either a single TADF material or two materials deposited simultaneously (cohost) to yield record performance OLEDs.

The very first commercial OLED product was introduced by Pioneer Corporation in 1997, which was a passive matrix organic light-emitting diode (PMOLED) display for car audio screens. However, it was not until a decade later in 2007 that Samsung Mobile Display introduced the first commercial AMOLED display, which remains to be the screen of choice for portable smart phones and tablets. As a general light source, the first white OLED was reported in 1995 by mixing emitters of the three primary colors into a single OLED device to produce white light.[4] The panel efficiency of white OLEDs has just reached ~130 lm/W at 1000 cd/m^2, which well exceeds the performance of a standard fluorescent tube (~70 lm/W). Although the performance of OLED panels has been greatly improved, there is still much room (nearly 120 lm/W) for improvement in efficacy, when considering a theoretical limit of ~250 lm/W. Currently, OLED for lighting remains an active target globally with key challenges revolving around extending the lifetime of the blue emitters, reducing fabrication steps (and cost), improving light extraction, and increasing device stability for high-brightness operations under a continuous electrical drive on both flexible and rigid substrates.

Fundamentally, an OLED is an EL device made up of layers of functional organic materials with tens of nanometers thickness that are stacked between two electrodes, an anode and a cathode. In order for light to escape out of the device, one of the electrodes must be transparent. To avoid water vapor and oxygen exposure that react with the organics to form light-quenching centers, these organic layers are typically deposited by thermal evaporation under ultrahigh vacuum (<1 × 10^{-7} Torr) environment. A variety of substrates including glass, plastic, Si, or even stainless steel could be used for the OLEDs to be deposited on as long as they are sufficiently smooth (under ~30 nm in roughness).

Figure 1.2 shows an illustration of the OLED working principle. Under an applied bias voltage that induces an electric field, holes (electrons) are injected from the anode (cathode) and migrated through the hole (electron) transport layer to the emissive layer (EML) at the center. Since organic molecules are held together by weak van der Waals forces, the migration of charges is a random, molecule-to-molecule, hopping process. The applied voltage is large enough such that the difference in quasi-Fermi levels formed between the two electrodes exceeds the energy gap of the host (typically >2.5 eV) in order to be able to supply charge carriers with sufficient energy and density into the host layer. Once inside the EML, electrons and holes pair up to form tightly bound pairs called excitons due to Coulomb interaction.[5] Here, the EML typically consists of a host–dopant combination where the dopant has a lower energy corresponding to the visible spectral range and is incorporated in small fractions into the host matrix. In this way, once high-energy excitons are formed in the host, they are able to reach thermodynamic equilibrium by performing excitonic energy transfer, either of a Förster[6] or Dexter[7] type, to form lower-energy excitons on the dopant molecules facilitated by a close spatial proximity and a strong spectral overlap between the host (donor) and the dopant molecules (acceptor). These initially formed excitons on the dopant molecules then quickly relax to the lowest singlet (S_1) and triplet (T_1) states following Kasha's rule[5] as a result of the much faster internal conversion (IC) rate than the intersystem crossing (ISC) rate, before finally releasing the energy either radiatively to produce vibrant, visible colors, or nonradiatively as heat via molecular vibrations and other quenching processes. After radiative emission, the absorption by the same type of dopant or other lower-energy dopants in the device is minimal since the dopants are present only in small fractions within the relatively thin EML layer. In general, the nonradiative relaxation rate of the lowest-energy excitons on the dopant molecules follows the energy gap law:[8]

$$k_{nr} = 10^{13}\, e^{-\alpha E_g}, \qquad (1.1)$$

where α is a proportionality constant depending on the nature of the molecule, and E_g represents the energy gap as determined from the dopant molecule's lowest singlet-exciton energy level. This implies that the lower the emission energy (longer emission wavelength), the higher the likelihood of nonradiative processes occurring, hence the less efficient the dopants are.

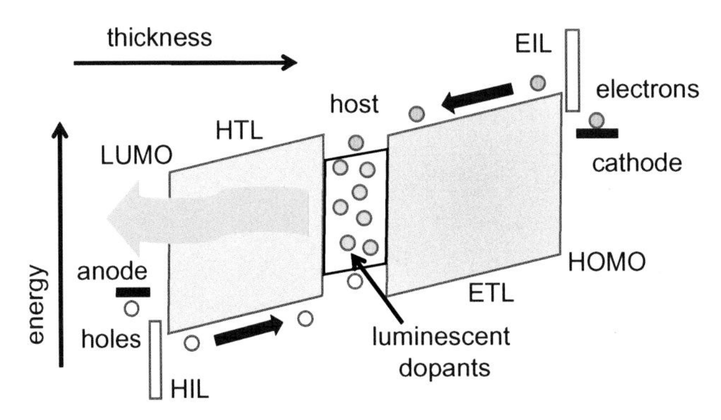

Figure 1.2 Illustration of the working principle of an OLED under forward electrical bias. LUMO and HOMO levels denote the lowest unoccupied molecular orbital and the highest occupied molecular orbital of the organics, respectively. HIL and EIL represent hole injection layer, and electron injection layer, respectively. HTL and ETL are hole transport and electron transport layers, respectively. Solid yellow arrow represents light directed out of the device.

An example of a simple, yet effective OLED device is illustrated in Fig. 1.3.[9] Here, 2,2′,2″-(1,3,5-benzinetriyl)-tris(1-phenyl-1-H-benzimidazole) (TPBi) is used as the electron transport layer (ETL), and 4,4′-bis(carbazol-9-yl)biphenyl (CBP) is employed as both the host and hole transport layer (HTL). A well-known green phosphorescent guest molecule, $Ir(ppy)_2(acac)$ [bis(2-phenylpyridine) (acetylacetonate) iridium(III)],[10] is doped in ~4 wt.% into the CBP host with a thickness of 15 nm to form the EML. The hole and electron injection layers (~1 nm thick) are MoO_3 and LiF, respectively. The cathode employed is Al with high optical reflectivity in the visible range. The transparent conducting anode is an indium tin oxide (ITO) film (~50–120 nm) prepatterned on a glass substrate. At the time of this writing, the best green OLED on glass substrates without employing additional light extraction techniques exhibits a maximum external quantum efficiency of 32.3%,[11] which corresponds to an efficacy of 142.5 lm/W.

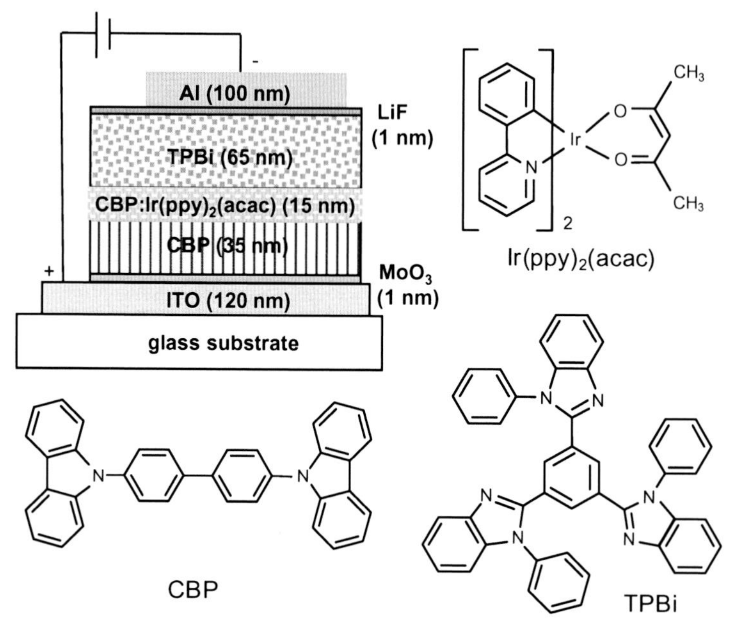

Figure 1.3 A simple green bottom-emission OLED utilizing a well-known phosphorescent green dopant.

Chapter 2

OLED Working Principles

Before introducing the major types of efficient organic light-emitting diode (OLED) device architectures, it is appropriate to first outline standard device performance evaluation methods, introduce four major classes of dopant materials, and dwell deeper into the fundamental principles regarding exciton physics (or *excitonics*) that govern the operation of an OLED.

2.1 Performance Evaluation

In characterizing organic light-emitting diode (OLED) performance, the most common parameter is the external quantum efficiency (η_{EQE}), which is defined as the number of photons generated per number of charge carriers injected. It can be described as follows:[12]

$$\eta_{EQE} = \eta_{oc}\,\eta_{IQE}, \quad (2.1)$$

$$= \gamma\eta_{oc}\,\eta_{e-p}, \quad (2.2)$$

$$= \gamma\eta_{oc}\,\chi\phi_{PL}, \quad (2.3)$$

where η_{oc} is the optical out-coupling efficiency, γ is a charge carrier balance (CCB) factor, η_{e-p} is the exciton to photon conversion efficiency, and η_{IQE} represents the internal quantum efficiency. Here, η_{e-p} can also be represented by the product of χ, the fraction of emissive excitons received from the host or directly trapped by

Efficient Organic Light-Emitting Diodes (OLEDs)
Yi-Lu Chang

ISBN 978-981-4613-80-4 (Hardcover), 978-981-4613-81-1 (eBook)
www.panstanford.com

the emitter chosen, and ϕ_{PL}, the luminescence quantum yield of the emitter. Due a significant refractive index mismatch among the different materials used in the device, including the indium tin oxide (ITO) anode, glass substrate, and organic layers, a considerable amount of light is trapped inside the OLED from total internal optical reflections.[13] This results in a maximum light out-coupling efficiency of ~0.20–0.30 for standard OLEDs on glass substrates with standard ITO anodes. For a fluorescent emitter, χ is ~0.25. For a phosphorescent emitter or the recently introduced thermally activated delayed fluorescence (TADF) emitter, the maximum χ could reach unity. In the case of a well-optimized device having perfectly matched electron and hole currents, γ could also reach near unity. For an efficient phosphorescent emitter ($\phi_{PL} \approx 1$) such as $Ir(ppy)_2(acac)$, together with a highly compatible host such as CBP, $\eta_{e\text{-}p}$ could reach unity as well. The bottleneck in raising overall device efficiency is then the optical out-coupling or light extraction. Methods to improve this out-coupling efficiency are outlined in Chapter 9.

While η_{EQE} measures the number of photons extracted to air divided by the number of injected charges, current efficiency, η_{CE}, and power efficiency (or efficacy), η_{PE}, are two other useful parameters which are both photometric quantities that also take into consideration the photosensitivity of human eyes.

The current efficiency is calculated using a measured luminance $L_0{}^{o}$ in the forward direction together with a measured current density J_{meas} passing through the device:

$$\eta_{CE} = \frac{L_0{}^{o}}{J_{meas}} \text{ [cd/A]}, \tag{2.4}$$

The power efficiency or efficacy is then computed using the operating voltage at the corresponding current density, $V(J_{meas})$, as follows:

$$\eta_{PE} = \eta_{CE} \frac{f_D \pi}{V(J_{meas})} \text{ [lm/W]}, \tag{2.5}$$

with

$$f_D = \frac{1}{\pi I_0} \int_0^{\pi/2} \int_{-\pi}^{+\pi} I(\theta,\phi) \sin\theta \, d\phi \, d\theta, \tag{2.6}$$

where f_D is the angular distribution of the emitted light intensity $I(\theta,\phi)$ in the forward hemisphere as a function of the azimuthal (θ) and polar (ϕ) angles. I_0 denotes the light intensity measured in the forward direction perpendicular to the emitting surface. In general, the emission spectra of OLED may be altered at different angles of view, which will be discussed in detail in Chapter 7.

The external quantum efficiency, η_{EQE}, can be acquired by[14]

$$\eta_{EQE} = \eta_{CE}\frac{f_D \pi e}{K_r E_{ph}}\left[\frac{\%}{100}\right], \tag{2.7}$$

where E_{ph} is the average photon energy of the electroluminescent (EL) spectrum and e is the electron charge. K_r is the luminous efficacy of radiation, which can be calculated by

$$K_r = \frac{\int_{380\,nm}^{780\,nm} \Phi_r(\lambda)V(\lambda)d\lambda}{\int_0^{\infty} \Phi_r(\lambda)d\lambda}\ [lm/W], \tag{2.8}$$

where $V(\lambda)$ is the weighting function that takes into consideration the photosensitivity of human eyes and Φ_r is the radiant flux. In essence, K_r quantifies lumen per watt for a given spectrum, thereby also representing the theoretical limit in power efficiency of a particular light source, assuming no optical and electrical losses. It is important to note that the angular distribution f_D has to be properly measured using an integrating sphere in order to obtain both η_{EQE} and η_{PE} accurately.

2.2 Emitter Classifications

For electrically excited organic molecules, the energy may be released radiatively either through a fluorescent or a phosphorescent process. The fundamental mechanisms of fluorescence and phosphorescence are illustrated in Fig. 2.1a. According to spin statistics in quantum mechanics,[5] electrically excited excitons in organic molecules are classified as singlet (S) and triplet states (T), with an electronic state density ratio of 1:3. In a fluorescent molecule, the triplet states are nonemissive; hence only a quarter of the total excitons generated may contribute to light emission from its lowest singlet state (S_1). Such singlet energy radiative relaxation happens on a relatively fast time scale of $\sim 10^{-9}$ s.[5] Conversely, in a phosphorescent molecule, the molecule is attached to a heavy metal atom such as Ir or Pt, which can induce a spin–orbit coupling effect, leading to a rapid exciton energy transfer from the singlet to the triplet state (intersystem crossing [ISC]), as well as allowing for a spin-flip that enables a triplet state to relax to the ground state radiatively.[5] The energy relaxation time of the triplet states is in the order of $>10^{-6}$ s.[2] Here, the strength of spin–orbit coupling is critical and it is directly proportional to the fourth power of the atomic number of the metal; hence the heavier the metal, the stronger the spin–orbit coupling and the higher the emission efficiency.[15] Essentially, these processes lead to potentially 100% of the electrically generated excitons contributing to light emission. Hence, the use of a phosphorescent emitter yields a fourfold enhancement in light emission efficiency (see Fig. 2.1b) over that of a fluorescent emitter. On the basis of the three orders of magnitude difference in emission decay time scale between fluorescent and phosphorescent emitters, it can be understood that singlet radiative emission leads to much less accumulation, whereas triplets accumulates quickly under high current density.[16] This inevitably results in more severe exciton–exciton quenching[16] and hence a faster roll-off in efficiency,[17] as shown in Fig. 2.1b.

Figure 2.2 shows the chemical structures of well-known green, red, and blue phosphorescent emitters.[18,19] In terms of the primary colors suitable for commercial applications, green and red phosphorescent emitters are adequate in terms of both efficiency and lifetime for most display applications. It remains a challenge, however, to obtain a stable blue phosphorescent OLED because of the fact that the energy required to excite the blue emitter is close to that of the dissociation energy of the common C–C and C–N chemical bonds in the organic complex.[20] In addition, it has recently been found that by pushing the metal-to-ligand charge transfer excited state of these metal–organic molecules into the higher energy blue region, a nonradiative pathway is introduced via the metal *d*-orbitals, which makes the molecule thermally and photochemically unstable.[21] Additionally, to excite the high-energy blue emitters, even-higher-energy (or wider-energy-gap) host materials have to be electrically excited first. This further leads to host molecule instability issues (e.g., aggregation and fragmentation), as will be discussed in Chapter 10.

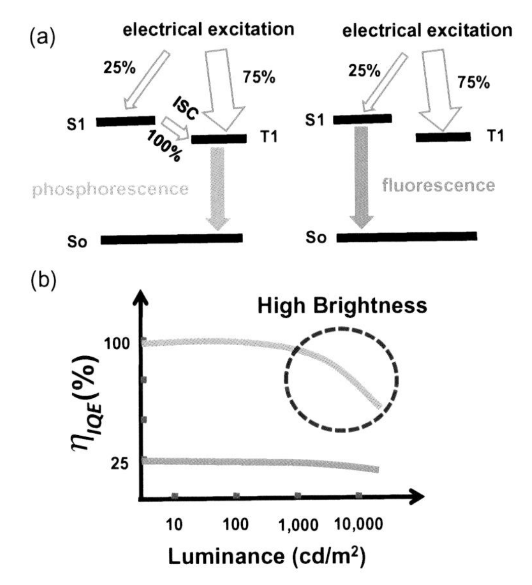

Figure 2.1 (a) Illustrations of fluorescence and phosphorescence processes as well as (b) their corresponding internal quantum efficiency as a function of luminance. Open and solid arrows indicate nonradiative and radiative energy transitions, respectively. S_1, T_1, and S_0 represent energy states from the lowest singlet, triplet, and ground state, respectively. ISC stands for intersystem crossing.

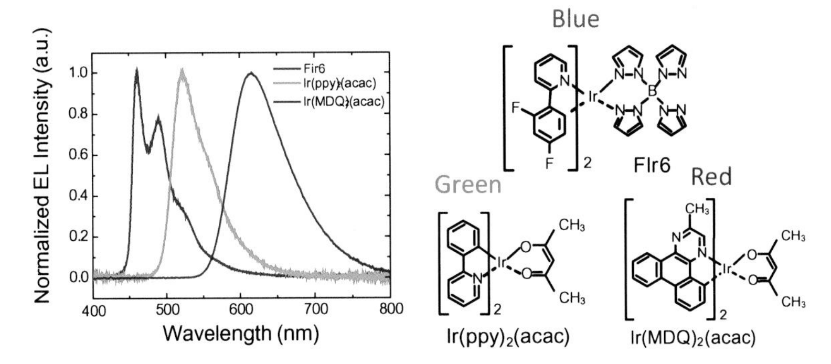

Figure 2.2 EL spectra and chemical structures of standard Ir-based phosphorescent emitters. Bis(4,6-difluorophenylpyridinato) tetrakis(1-pyrazolyl)-borate iridium(III) (FIr6) is the blue emitter, $Ir(ppy)_2(acac)$ is the green emitter, and bis(2-methyldibenzo[f,h]quinoxaline) (acetylacetonate) iridium(III) ($Ir(MDQ)_2(acac)$) is the red emitter.

One unique type of fluorescent emitter is based on triplet–triplet annihilation (TTA) or a triplet fusion process to produce additional singlets for enhanced fluorescent emission as expressed in Eq. 2.10. Here, η_r is the fraction of singlets produced by electron–hole recombination and η_{TTA} is the fraction of extra singlets generated from two triplets annihilating with one another. This concept exploits the fact that triplets in fluorescent emitters cannot release the energy radiatively, and therefore accumulate rather quickly. By way of triplet fusion, these accumulated triplets can instead be utilized to form high-energy singlets that quickly relax to the lowest excited singlet state to be radiatively emitted upon relaxation.[22] This can surpass the traditional bottleneck of 25% singlet generation upon electrical excitation to provide ~62% of total singlets,[21] leading to a theoretical η_{EQE} of as high as 12.5% for fluorescence instead of 5% conventionally.

$$\eta_{EQE} = \gamma\eta_{oc}\chi\phi_{PL}, \tag{2.9}$$

$$= \gamma\eta_{oc}(\eta_r + \eta_{TTA})\phi_{PL}, \tag{2.10}$$

$$= (1.0)(0.2)\left(0.25 + \frac{0.75}{2}\right)(1.0) = 12.5\%$$

This type of emitters has recently being industrially utilized as the blue emitter of choice since phosphorescent blue emitters remain subpar in terms of stability and lifetime.

Recently, another class of fluorescent emitters that displays efficiencies close to that of a phosphorescent emitter has been reported.[3] The light-emitting process of this type of molecules is based on TADF, shown in Fig. 2.3. Here, the energy-level difference between the singlet and triplet states of the emitter is made sufficiently close (<100 meV) such that with a small thermal energy input during device operation (at room temperature), the slightly lower-energy triplet excitons could transfer to the higher singlet energy level (reverse intersystem crossing [RISC]) efficiently, resulting in nearly 100% exciton fluorescence from the singlet states.[3] Such delayed fluorescence process occurs on a time scale of ~10^{-6} s, which is comparable to that of the radiative process in phosphorescence. The key to achieving a small singlet and triplet energy difference (exchange energy) is to separate the electron donor and acceptor moieties of the molecule thereby minimizing the wavefunction orbital overlap between the molecule's highest occupied molecular orbital (HOMO) and lowest unoccupied molecular orbital (LUMO) levels, while preserving a high dipole oscillator strength for high emissive yield.[23]

The use of this new class of molecules could avoid the use of phosphorescent emitters that contain expensive, scarce rare-earth metals such as Ir and Pt. However, the design of TADF molecules is not trivial and the synthesis procedures may still require expensive catalysts such as Pd and other rare earth metals. In addition, it remains to be seen if the lifetime could surpass those of state-of-the-art fluorescent and phosphorescent emitter-based OLEDs since the TTA process may still be significant owing to the possible accumulation of nonradiative triplet states. Interestingly, as will be discussed in detail later in Chapter 5, materials based on this concept of TADF could also act as ideal hosts.[24]

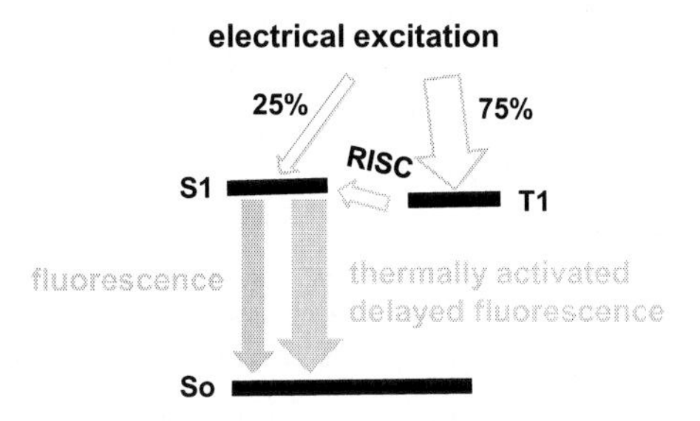

Figure 2.3 Illustration of thermally activated delayed fluorescence process. Open and solid arrows indicate nonradiative and radiative energy transitions, respectively. RISC stands for reverse intersystem crossing.

2.3 Excitonics

In organic semiconductors, weak van der Waals interactions holding each molecule together induce a weak dielectric screening of the Coulomb interactions due to randomly oriented polarizations. This is reflected by a small dielectric constant ε_r of ~3–4, that leads to a large exciton binding energy (0.3–1 eV) and tightly bound electron–hole pairs that are either localized on a single molecule (0.5–1 nm) called Frenkel excitons or on adjacent molecules called charge transfer excitons.[18] Such localized electron–hole pairs also leads to a strong electron–hole wavefunction overlap that introduces a large exchange energy (0.1–1 eV), which separates the singlet and the triplet state energies apart. From Pauli's exclusion principle, the singlet excitons have antisymmetric spin wavefunctions and are therefore spatially bound closer together with a higher energy, whereas the triplet excitons have symmetric spin wavefunctions such that they are spatially further apart with a lower energy due to strong spin–spin interactions. These are in stark contrast to inorganic semiconductors that are formed by strong ionic and covalent bonds which exhibits a strong dielectric screening of the Coulomb interactions due to well-ordered polarizations. As a result, large dielectric constants ε_r in the range of ~11–16 are observed. In effect, the exciton binding energies are in general small (14.7 meV for Si, 4.7 meV for GaAs, and 2.7 meV for Ge) and the electron–hole pairs are loosely bound (4–10 nm) or also called Wannier excitons. Such loosely bound electron–hole pairs suggests little electron–hole wavefunction overlap, leading to nearly zero exchange energy (on the order of a few meV), and hence there is no need to differentiate between a singlet and a triplet exciton (simply called an exciton).[25]

In addition to radiative recombination of charge carriers in which excitons are formed as excited states in organic molecules prior to releasing the energy radiatively, two important types of nonradiative energy transfer mechanisms, namely Förster type and Dexter type, are also vital to OLED operation. These mechanisms between host and guest are schematically shown in Fig. 2.4.

Here, a molecule involved in an excitonic energy transfer process is referred to either as a donor D or as an acceptor A, depending on whether the molecule donates or accepts energy, respectively. Furthermore, the multiplicities of the excitons are denoted with preceding superscripts, 1 (^{1}D and ^{1}A) or 3 (^{3}D and ^{3}A) for singlets and triplets, respectively, and species in the excited states are marked with asterisks (D* and A*).

For the Förster transfer a significant overlap of the emission spectrum of the host matrix and the absorption spectrum of the dopant is crucial.[5] The main driving force of the Förster energy transfer mechanism is a dipole–dipole interaction between the dipole transition moments $\mathbf{M}_1$ and $\mathbf{M}_2$ (Fig. 2.4a). Schematically this can be shown by simultaneous electron (hole) "jumps" in HOMO–LUMO configurations (denoted by dashed arrows in the middle of Fig. 2.4a).

The second type of energy transfer is the short-range mechanism of Dexter (Fig. 2.4b). Its driving force is the exchange interaction expressed by the exchange integral.[5] In the simplified diagram, the exchange integral is essentially a measure of the electron exchange rate between the HOMO and LUMO of the donor and acceptor that occurs concurrently, as shown by dashed arrows. As with Förster transfer, a considerable overlap of the emission spectrum of the host matrix and the absorption spectrum of the dopant is necessary.

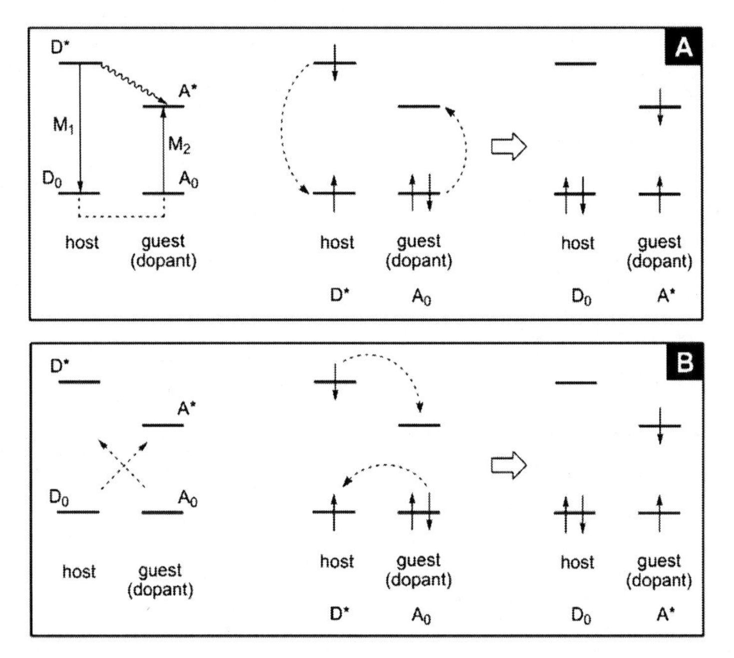

Figure 2.4 Simplified charge transfer diagrams for (a) Förster and (b) Dexter transfer mechanisms. Reprinted with permission from Ref. [5]. Copyright 2014, Royal Society of Chemistry.

The nonradiative energy transfer rate for both mechanisms is directly proportional to the spectral overlap of the donor emission band $I_D(\nu)$ and the acceptor absorption band $\alpha(\nu)$. This is described by the spectral overlap integral J as follows:

$$J = \int_0^\infty I_D(\nu)\alpha(\nu)d\nu, \tag{2.11}$$

where $I_D(\nu)$ and $\alpha(\nu)$ are normalized intensities.

In the case of Förster energy transfer[26] that is driven by Coulomb interaction–induced dipole–dipole coupling, the rate constant represented by the most dominant dipole–dipole interaction can be approximated as[27]

$$k_{\mathrm{F}} = k_0 \frac{9(\ln 10)\kappa^2 \phi_{\mathrm{D}}}{128\pi^5 N_{\mathrm{A}} n^4} \cdot J \cdot \frac{1}{R_{\mathrm{DA}}^6} = k_0 \left[\frac{R_{\mathrm{o}}}{R_{\mathrm{DA}}} \right]^6, \tag{2.12}$$

where k_0 is the rate constant of the excited donor without the presence of an acceptor, κ denotes the orientation factor, N_{A} is the Avogadro's number, n is the refractive index of the medium, ϕ_{D} is the luminescence quantum yield of the donor emission, R_{DA} is the intermolecular distance between a donor and an acceptor, and R_0 is known as the Förster radius. Under this framework of Förster transfer, the following processes are possible:[14]

$$^1\mathrm{D}^* + {}^1\mathrm{A} \rightarrow {}^1\mathrm{D} + {}^1\mathrm{A}^* \qquad [2.1]$$

$$^1\mathrm{D}^* + {}^3\mathrm{A} \rightarrow {}^1\mathrm{D} + {}^3\mathrm{A}^* \qquad [2.2]$$

These processes imply that a donor molecule with excited singlet states could transfer its energy to either the singlet or triplet states of an acceptor by Förster transfer. This is thus a necessary energy transfer mechanism between a standard host and a guest emitter in an OLED. Such proficient singlet–singlet Förster transfer also sets the limit on the doping concentration of typical fluorescent dopants in a host to be ~1% or less in order to prevent significant emitter self-quenching (repeated self-absorption and re-emission, while losing more energy through nonradiative paths each time). This is also why the most efficient OLEDs involve a combination of a host matrix and an emissive dopant as the emissive species cannot exhibits high concentration like a host matrix, while preserving high radiative emission efficiency. In addition, if phosphorescent donor molecules are involved, two additional processes are allowed due to an enhanced triplet recombination induced by a strong spin–orbit coupling that facilitates a spin-flip of the triplet states:[21]

$$^3\mathrm{D}^* + {}^1\mathrm{A} \rightarrow {}^1\mathrm{D} + {}^1\mathrm{A}^* \qquad [2.3]$$

$$^3\mathrm{D}^* + {}^3\mathrm{A} \rightarrow {}^1\mathrm{D} + {}^3\mathrm{A}^* \qquad [2.4]$$

These processes become important when two or more phosphorescent dopants are incorporated in the same host as in the case for multicolor or white emission OLEDs. In this case, significant energy transfer can take place between the high- and low-energy phosphorescent dopants by way of Förster-type mechanism, leading to considerable quenching of the higher-energy dopant emission by the lower-energy ones.[28] Such energy transfer may occur even at low emitter doping concentrations because Förster energy transfer can be very efficient even at a long range of ~10 nm[6] that is considerably larger than the size of individual organic molecules.[29] Process [2.4] can also describe concentration-dependent phosphorescent emitter self-quenching or repeated self-absorption and re-emission, thereby losing more energy nonradiatively in each cycle.

In contrast, Dexter energy transfer[7] comes from electron exchange interactions that require significant orbital overlap between D and A. The Dexter transfer rate constant is expressed as[21]

$$k_{\mathrm{D}} = \frac{2\pi}{\hbar} K^2 \cdot J \cdot e^{-2R_{\mathrm{DA}}/L}, \tag{2.13}$$

where K denotes a constant with units of energy and L represents the sum of van der Waals radius. Here, the exponential dependence on the intermolecular distance R_{DA} indicates the quantum mechanical nature of closely bound electrons that have sufficient wavefunction overlap to facilitate such Dexter exchange process. Typically, Dexter transfer distance is only up to ~2 nm.

In addition, Dexter-type energy exchange follows the Wigner–Witmer spin conservation rules, which requires the total spin configuration to be conserved throughout the process. The resulting energy transfer processes are[21]

$${}^1\mathrm{D}^* + {}^1\mathrm{A} \rightarrow {}^1\mathrm{D} + {}^1\mathrm{A}^* \qquad [2.5]$$

$${}^3\mathrm{D}^* + {}^1\mathrm{A} \rightarrow {}^1\mathrm{D} + {}^3\mathrm{A}^* \qquad [2.6]$$

$${}^3\mathrm{D}^* + {}^3\mathrm{A}^* \rightarrow {}^1\mathrm{D} + {}^1\mathrm{A}^* \qquad [2.7]$$

Even though singlet to singlet energy transition is possible from Dexter exchange interactions as indicated in Process [2.5], this transition is most likely to occur by highly efficient Förster transfer in Process [2.1]. Process [2.6] describes the triplet migration process or "hopping" transport through closely touching organic host molecules. Basically, triplets in the host will continue to migrate until a suitable guest molecule is found whereby energy is transferred to the triplet state of the guest by the same process. Typically, triplets in an organic semiconductor will diffuse a fairly long distance (~100 nm, on the order of entire device length) without radiative emission to the ground states since such transition requires a spin-flip, which is not possible without the help of heavy metal-induced spin–orbit coupling effect. Process [2.7] implies that two excited triplet states may react and generate two singlet states, one in the ground state and one in the excited state. This process is also known as TTA,[16] which may lead to phosphorescent OLED efficiency roll-off under high driving voltages or high current densities when the device is completely filled with excited triplet states (see Fig. 2.1b). Interestingly, this process is also called triplet fusion (see Eq. 2.10) when the resulting excited singlet state relaxes to the lowest singlet state to emit radiatively in a desired wavelength.

In terms of exciton migration throughout organic matrix during device operation, both Förster (mainly singlet to singlet) and Dexter (mostly triplet to triplet) energy transfers may be at play.[30,31] For exciton diffusion, because there is no net charge, the driving force behind exciton movement is a gradient in exciton concentration, $\nabla n(\mathbf{r},t)$. This gradient triggers a series of uncorrelated hopping processes from one molecule to another in a random-walk fashion. Such particle diffusion phenomenon can be modelled by Fick's second law as follows:[32]

$$\frac{\partial n(\mathbf{r},t)}{\partial t} = G(\mathbf{r},t) - \frac{n(\mathbf{r},t)}{\tau} + D\nabla^2 n(\mathbf{r},t), \tag{2.14}$$

where $G(\mathbf{r},t)$ denotes exciton generation, D is the diffusion constant, and τ represents the exciton lifetime. Moreover, triplet exciton diffusion may also be dependent on the bipolar transport property of the organic material since simultaneous electron and hole exchange is necessary to facilitate Dexter energy transfer. This means triplet excitons exhibit longer diffusion lengths and may be more dispersed in a bipolar material.

Under forward electrical excitation, excitons are generated typically in a close proximity to an interface between two strongly opposite charge–transporting organic layers such that the width of the exciton generation zone is considerably narrower than the thickness of total device organic stack. In this case, it is viable to model exciton generation zone as a delta function, that is, $G(x,t)$ =

$G \cdot \delta(x = x_0, t)$. Under this condition, one could obtain the steady-state ($\partial n/\partial t = 0$) solution of Fick's second law as follows:[33]

$$n(x) = n_0 \cdot e^{-x/L_x}, \; L_x = \sqrt{D\tau}, \tag{2.15}$$

where L_x represents the diffusion length and n_0 denotes the exciton density at the interface.

On the basis of these insights regarding exciton physics, researchers have only recently acquired considerable success in reducing the electron–hole wavefunction overlap on a single organic molecule by molecular design (forming a Frenkel exciton) or by way of mixing two different organic molecules (forming a charge-transfer exciton) in order to lower the exciton binding energy and hence minimizing the exchange energy such that the singlet and triplet exciton energy levels of the organic material become very close. This effectively reduces the amount of triplet excitons by allowing an efficient RISC process to the singlet states given a small thermal energy input. When it takes place in an emitter, it reduces the amount of the nonradiative triplet excitons in typical organic molecules that are without heavy metal-induced spin–orbit coupling effect. When it occurs in a host, it promotes the dominance of singlet-to-singlet Förster-type long-range energy transfer over triplet-to-triplet Dexter-type short-range energy transfer to the dopants. This means that when TADF hosts are doped with phosphorescent emitters, an enhanced singlet-to-singlet, host-to-dopant energy transfer will take place effectively at lower doping concentrations (less concentration induced self-quenching among dopant excitons) and require lower turn on voltages, thereby improving the emission efficiency, while lowering material cost. When TADF hosts are doped with fluorescent emitters, it would still be possible even for the fluorescent emitters to harvest nearly 100% of the excitons generated in the host during OLED operation owing to efficient singlet-to-singlet Förster energy transfer from the TADF host, which has the ability to convert nearly all of its initially generated triplet excitons to singlet excitons by RISC. Indeed, these TADF-type dopants and hosts are beginning to burgeon in the research field, and they are projected to aid or potentially replace the phosphorescent and fluorescent technologies that are currently dominating the OLED industry. These novel approaches to achieve ultrahigh-performance devices will be discussed in detail in Chapter 5.

Chapter 3

Charge Carrier Injection and Transport

In this chapter, two critical components of an organic light-emitting diode (OLED) device that determine the current injection process, namely, anode hole injection layer and cathode electron injection layer, are first presented in detail. An effective carrier injection layer ensures an Ohmic contact that minimizes the amount of undesirable charge accumulation at either electrode, which would contribute to a waste of electrical energy input. The transport across barriers formed by different organic materials will then be discussed. Figure 3.1 illustrates the sequential processes that carriers experience in an OLED including injection, migration, collision, and recombination.

Charge carrier transport and injection in an OLED are two key factors that dictate the performance of a device. For example, a poor charge injection at the electrode and/or a slow charge transport in the organic layers will result in a higher driving voltage, which means a higher power is required to generate sufficient amount of excitons in the host to produce a desired brightness. On the cathode end, a low-work-function metal such as Al in conjunction with a dielectric electron injection layer such as LiF, Liq, or $CsCO_3$ are used to achieve sufficient electron injection. On the anode end, a transparent conducting material such as indium tin oxide (ITO) combined with a high-work-function hole injection layer such as MoO_3,[34] poly(3,4-ethylenedioxythiophene):poly(4-styrenesulfonate) (PEDOT:PSS)[35] and 1,4,5,8,9,11-hexa-azatriphenylene hexa-carbonitrile (HATCN)[36]

Efficient Organic Light-Emitting Diodes (OLEDs)
Yi-Lu Chang

ISBN 978-981-4613-80-4 (Hardcover), 978-981-4613-81-1 (eBook)
www.panstanford.com

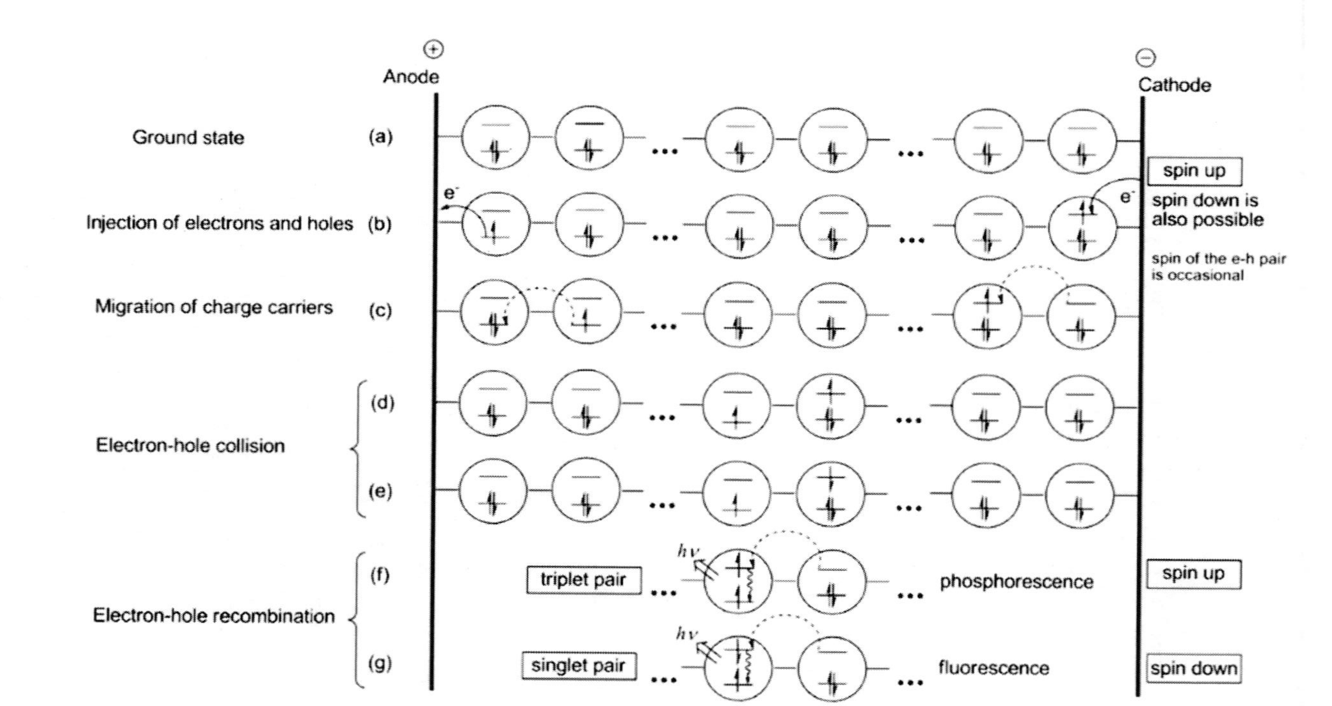

Figure 3.1 Simplified charge carrier energy state diagram at different stages of OLED operation. Reproduced with permission from Ref. [5]. Copyright 2014, Royal Society of Chemistry.

is used to ensure effective hole injection. The energy-level diagram at an ITO/organic interface is shown in Fig. 3.2, which highlights the need for a thin layer of MoO_3 to lower the injection barrier. Recently, it was found that chlorinated ITO surface could also increase the work function sufficiently high for effective hole injection.[37]

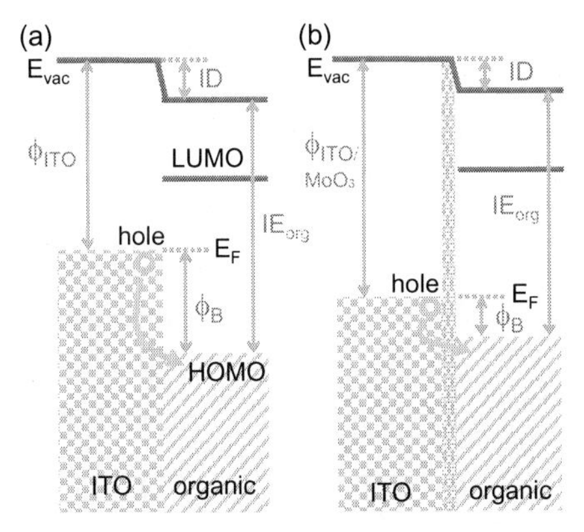

Figure 3.2 Simplified energy-level diagram across an organic–ITO anode interface without (a) and with (b) a hole injection layer of MoO_3. Here, ϕ_{ITO} denotes the work function of ITO, *ID* denotes the interface dipole, and IE_{org} stands for ionization energy of the organics. It can be seen that the injection barrier ϕ_B is reduced with the presence of MoO_3.

In general, current transport across an OLED may be classified into two different types. The first one assumes an Ohmic contact is present at the charge injection electrode/organic interface so that there is no electric field drop across the interface. In this case, the applied external voltage could be entirely used to drive charge carrier transport across the organic layer. The current–voltage characteristics in this case are called a space charge–limited current (SCLC), which can be described by Mott–Gurney law:

$$J = \frac{9}{8}\varepsilon_0 \varepsilon \mu \frac{V^2}{d^3}, \tag{3.1}$$

where V represents the applied voltage, d is the thickness of the organic film, μ denotes the charge carrier mobility, ε_0 is the permittivity of free space, and ε denotes the dielectric constant of the organics. Because most organic semiconductors have a field-dependent mobility, one can further express μ using the Poole–

Frenkel type of field dependence:

$$\mu(F) = \mu_0 e^{\beta\sqrt{F}}, \tag{3.2}$$

where F is the electric field strength. An approximation to the SCLC can be expressed as follows:[38]

$$J_{\text{SCLC}} = \frac{9}{8}\varepsilon_0\varepsilon\mu_0\frac{V^2}{d^3}e^{\beta\sqrt{V/d}} \tag{3.3}$$

Generally, if a chosen electrode/organic contact is able to inject enough current to sustain this above current density at a given voltage, the contact may be considered Ohmic.

In the second case, there exists a considerable energy barrier at the metal–organic interface, similar to that of a Schottky contact at a metal/semiconductor interface. In this case, the current density is limited by the amount of charges injected from the electrode instead of carrier mobility in the organic layer. This is called injection-limited current (ILC), which has been mathematically modelled by Scott and Malliaras. Here, the ILC is based on a solution to the drift–diffusion equation for the injection into a wide band gap intrinsic semiconductor that was solved by Emtage and O'Dwyer. This model takes into account the equilibrium contributions to the current density from charge carrier recombination with their own image charge, which is similar to Langevin recombination of an electron–hole pair in the bulk material. In this case, the current density can be expressed as[39]

$$J_{\text{ILC}} = 4N_0\psi^2 e\mu F e^{-e\phi_{\text{B}}/k_{\text{B}}T} e^{f^{1/2}}, \tag{3.4}$$

where F is the applied electric field, μ is the field-dependent carrier mobility, N_0 denotes the density of states in the organic film, ϕ_{B} is the barrier height, k_{B} is the Boltzmann constant, e is the electron charge, T is the temperature, and ψ is a function of the reduced electric field $f = e^3F/4\pi\varepsilon k_{\text{B}}{}^2T^2$, defined by[39]

$$\psi = f^{-1} + f^{-1/2} - f^{-1}(1 + 2f^{1/2})^{1/2}. \tag{3.5}$$

This ILC model takes into account the injection barrier height between the metal and the organic, which can be independently measured using ultraviolet photoelectron spectroscopy (UPS).[40,41]

Chapter 4

Efficient Device Architectures

Three standard architectural designs to achieve high-efficiency organic light-emitting diode (OLED) devices, namely, exciton and carrier confinement, energy barrier minimization, and emissive layer (EML) expansion, are discussed in this chapter. These are also beneficial to constructing long lifetime, stable devices, which are even more critical for industrialization.

4.1 Exciton and Carrier Confinement

Two of the most effective ways to achieve high-efficiency organic light-emitting diode (OLED) devices are exciton and carrier confinements inside the emissive layer (EML) where emitters are incorporated.[42,43] By preventing electrons (holes) from the cathode (anode) from travelling past the host layer, more electron–hole pairs could be generated in the host and hence more excitons would be available to be delivered to the dopants. This is often done by inserting a thin blocking layer, or a layer of organic material that is particularly good at only one type of carrier transport, adjacent to the host. This would effectively shift the exciton formation interface away from the host/transport layer interface to the host/blocking layer interface. Moreover, once excitons are formed inside the EML, they could still diffuse away to the adjacent transport layers and

Efficient Organic Light-Emitting Diodes (OLEDs)
Yi-Lu Chang

ISBN 978-981-4613-80-4 (Hardcover), 978-981-4613-81-1 (eBook)
www.panstanford.com

not be harvested by the emitters. In this case, the main approach to circumvent it is to employ blocking layers (and ideally transport layers) with a higher triplet energy than that of both the host and the dopant. In addition, the host layer itself also needs to have a higher triplet energy level than that of the dopant to prevent dopant-to-host energy back transfer. This is called exciton and carrier confinement design. The reason triplet energy instead of singlet energy is considered is that it is always the lowest energy level of each material. An example of an OLED device exhibiting both exciton and carrier confinement is shown in Fig. 4.1. Here, an HTL of 2,2'-bis(*m*-di-*p*-tolylaminophenyl)-1,1'-biphenyl (3DTAPBP) and an ETL of 1,3-bis(3,5-dipyrid-3-yl-phenyl)benzene (BmPyPb) is used. Both of these transport layers exhibit dominant monopolar transport properties hence are also blocking layers of the opposite carrier type. This makes sure most electrons and holes meet in the host region at the center of the stack. More importantly, duel (hole and electron) host layers with high triplet energy are employed, namely, 4,4',4''-tri(*N*-carbazolyl)triphenylamine (TCTA) and 2,6-bix(3-(carbazol-9-yl)phenyl)pyridine (DCzPPy). In this design, the orange and blue dopants have triplet energies that are smaller than both hosts and transport materials such that the most efficient way to release the excitonic energy during electrical excitation is by radiative emission. This design has led to a high η_{EQE} of 25% for white OLEDs back in 2008. Since then, most high-performance OLEDs to date, especially blue and white OLEDs, follow these two design rules.

Additionally, regarding device lifetime extension and stability enhancement, it is important to keep high-energy charge carriers and excitons within a stable host material, instead of allowing them to diffuse away to react with or break chemical bonds in the adjacent transport layers and especially the thin and delicate charge injecting organic–metal interfaces at either electrode contact.

4.2 Energy Barrier Minimization

Another way to improve the device efficiency is to minimize the number of organic heterojunction interfaces that typically introduces energy barriers.[44] Figure 4.2 illustrates an energy-level diagram of an OLED with the position of charge carriers shown.

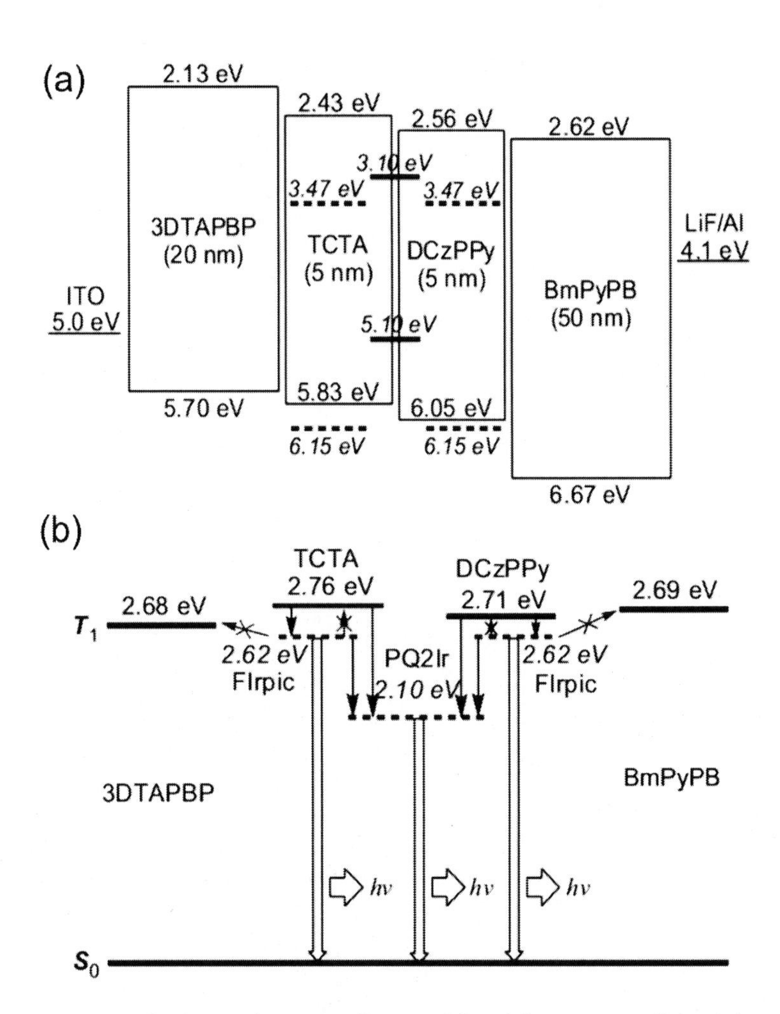

Figure 4.1 Device configuration diagram (a) and the corresponding triplet energy-level diagram (b) for a triplet- and exciton-confining design. Iridium(III) bis[2-(4,6-difluorophenyl)pyridinato-*N,C*$^{2'}$] (picolinato) (FIrpic) is a blue dopant and iridium(III) bis-(2-phenylquinoly-*N,C*$^{2'}$) dipivaloylmethane (PQ2Ir) is an orange dopant. Reproduced with permission from Ref. [42]. Copyright 2008, Wiley-VCH.

It can be seen that considerable amounts of charge carriers are stopped at the interface between two organic materials with large offsets in highest occupied molecular orbital (HOMO) or lowest unoccupied molecular orbital (LUMO) level, which form significant energy barriers. These energy barriers cause charge carriers to accumulate, thereby wasting electrical energy input since these charge carriers are not being transported to the host to contribute in exciton formation. In addition, such high density of charged carriers could also interact with and eliminate existing excitons in the vicinity of the interface through an exciton–polaron annihilation process, which will be discussed in detail in Chapter 10. One method to circumvent this issue is to reduce the complexity of the device by employing multifunctional organic materials in order to reduce the total number of different organic species in the device.

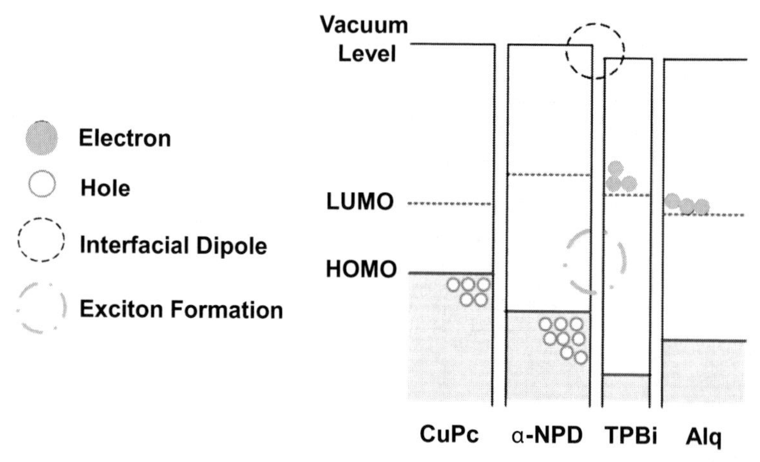

Figure 4.2 A device configuration diagram showing the position of carriers near the interfaces between two different organic layers. Reproduced with permission from Ref. [44]. Copyright 2010, American Institute of Physics.

Alternatively, one could also eliminate large energy barriers and reduce charge accumulation across an interface by applying gradient doping between each host and transport layers. This would facilitate charge transport through a number of shallow barriers rather than a single steep barrier across each interface. Such strategy has also proven to be critical in extending device lifetime.[45] However, this entails tedious fabrication procedures that could lead to device reproducibility and reliability issues, thereby significantly lowering production yield and increasing manufacturing cost.

4.3 Emissive Layer Expansion

A third approach to improve device efficiency is to expand the EML of the device. This involves a combination of exciton formation zone expansion and luminescent dopant layer extension. In a standard OLED device, the interface between two dominant, opposing (i.e., electron and hole conducting) transport layers would be where most electrons and holes meet and hence represents the initial interface of exciton formation. Starting from this interface, the exciton formation zone would extend toward both the cathode and the anode side, depending on the transport properties of the material. This is because exciton diffusion by Dexter transfer requires simultaneous exchange of electron and holes, hence excitons (especially triplets) will diffuse further inside the material with simultaneously decent electron and hole transport properties (i.e., ambipolar). Often, for complicated devices involving three or more organic layers, it may not be obvious where this initial exciton formation interface is. In this case, a simple technique presented by the author involves the insertion of a thin test transport layer with a smaller energy gap has proven to be successful in determining the correct interface.[46] If the test layer merely acts as a transport layer, the device characteristics shall not be altered significantly. Conversely, if the test layer significantly lowered the device performance, it means that the thin layer serves as an exciton sink, and the correct exciton formation interface would be identified.

With the knowledge that the exciton formation zone expands on both sides of the interface, it would be beneficial to incorporate the emitters on not only one side but on both sides of the interface.[46,47] This forms a dual-EML structure, which represents one way to expand the EML, thereby ensuring an emitter will be in close proximity to receive the high-energy exciton diffused toward either side of the interface.[47]

Additionally, it is important to point out another advantage of exciton formation zone expansion in OLEDs. Without such expansion, high density excitons formed in one narrow region would lead to exciton self-quenching processes that decrease the amount of available excitons. These quenching processes will be discussed in detail in Chapter 10. To expand the exciton formation zone further, researchers have performed codoping of two host materials having opposite transport properties to construct an ambipolar layer.[17,48] Such codoping may also be performed gradually (gradient doping)[47] in order to eliminate any sharp interfaces, hence reducing charge accumulation as mentioned previously. Figure 4.3 illustrates a summary of the various EML designs discussed thus far.[17] Alternatively, many groups have also performed molecular design to directly synthesize a single ambipolar material,[49,50] capable of both excellent hole and electron transport, as the host to simplify the device structure. Once ambipolar materials have been inserted, the EML could simply be expanded by doping the emitter through the entire thickness of the ambipolar layer. Such an ambipolar host material also aims at improving the charge balance in the EML of the device, which is a critical parameter for high-efficiency OLEDs (see Eq. 2.2).

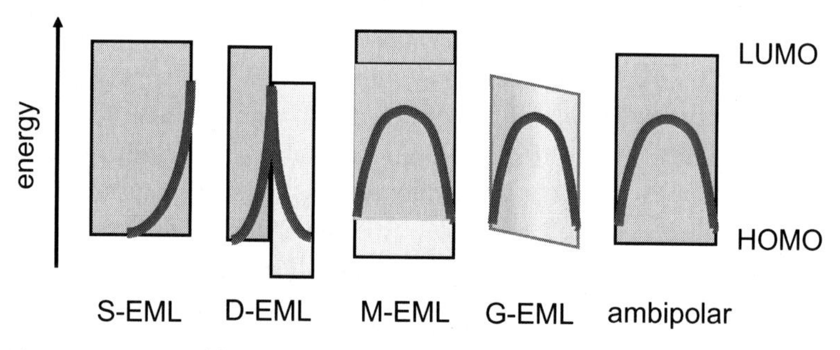

Figure 4.3 Possible EML host-mixing profiles to control charge balance and exciton formation zone. The black line indicates the projected spatial exciton density profile. S-EML, D-EML, M-EML, and G-EML denote single, double, mixed, and gradient emissive layer designs, respectively. Reproduced with permission from Ref. [17]. Copyright 2013, Wiley-VCH.

Aside from managing transport properties of the host in the EML and the region of dopant incorporation, the triplet exciton harvesting capability of a chosen host–dopant combination in a device is also critical in producing efficient devices, which will be discussed in the following chapter.

Chapter 5

Advanced Device Architectures: Exciton Harvesting

From Chapter 2 we learned that if the organic light-emitting diode (OLED) has only fluorescent emitters, only a quarter of the electrically generated excitons (singlets) will be harvested for light emission. This sets the limit on the internal quantum efficiency of the device to ~25%. To boost the performance toward 100% internal quantum efficiency, it is necessary to harvest the remaining 75% of the triplet excitons generated electrically in the device. In this chapter, we will discuss a number of advanced designs that can achieve this goal. First we will present a design involving triplet exciton harvesting and efficient energy transfer using phosphorescent emitters only. Additionally, it is known that triplet excitons in the host tend to spur various quenching processes either with themselves or with charged species, thereby accelerating device degradation. It is therefore beneficial to convert the bulk of the generated triplets into singlets in the device during operation by means of triplet to singlet up-conversion through reverse intersystem crossing (RISC). This is possible with the use of thermally activated delayed fluorescence (TADF) dopants and hosts to harvest triplet excitons and convert to singlets that can either be radiatively emitted directly or perform efficient energy transfer to other lower-energy dopants. Alternatively, this can be achieved using exciplex-forming cohost systems that not

Efficient Organic Light-Emitting Diodes (OLEDs)
Yi-Lu Chang

ISBN 978-981-4613-80-4 (Hardcover), 978-981-4613-81-1 (eBook)
www.panstanford.com

only converts triplets into singlets but also conveniently provides exciton confinement in the emissive layer (EML).

5.1 Exciton Harvesting via Phosphorescent Dopants

Phosphorescent dopants are already widely used in the display industry because of the ability to harvest both triplet and singlet excitons generated in the emissive layer (EML) to provide nearly 100% internal quantum efficiency. This exciton harvesting ability is critically dependent on the energy-level difference between the host and the phosphorescent dopant. Once a particularly suitable host–dopant match based on their energy level and energy gap difference is established, it is possible to incorporate other lower-energy dopants, either of a phosphorescent or a fluorescent type, in the same EML (intrazone[51, 52]) or in an adjacent EML (interzone[28]) in order to receive the harvested excitons by means of efficient excitonic energy transfer. When the lower-energy dopants receive additional excitons, a significant boost in their emission intensity is expected as illustrated in Fig. 5.1. This mechanism has been formulated by the author as[28]

$$\eta_{EQE} = \gamma\eta_{oc}\,\{i_A\chi_A\,\phi_{PL,A} + i_D\chi_D\,[\eta_{D-A}\phi_{PL,A} + (1 - \eta_{D-A})\phi_{PL,D}]\} \quad (5.1)$$

where η_{EQE}, η_{oc}, η_{E-P}, and η_{D-A} are device external quantum efficiency, out-coupling efficiency, exciton-to-photon conversion efficiency and energy transfer efficiency from donor to acceptor emitter, respectively. γ, ϕ_{PL}, and χ denote the charge balance factor, the absolute quantum yield of each emitter, and the fraction of emissive excitons trapped by each emitter in the device, respectively. Here, i is defined as the ideality factor accounting for the reduction in the fraction of emissive excitons received or trapped by each emitter with an EML thickness that deviates from the optimum thickness in a single-color device.

Figure 5.2a shows a device architecture where the green phosphorescent dopant performs exciton harvesting in the host CBP first, before carrying out efficient interzone exciton transfer to adjacent greenish-yellow phosphorescent dopant.[28] It can be seen that with the addition of sufficiently thick greenish-yellow EML, there is enough low-energy luminescent dopants to receive the harvested excitons from the high-energy harvesting dopant such that the energy transfer is nearly 100% (see Fig. 5.2b) and the emission becomes primarily coming from the greenish-yellow dopant (see Fig. 5.2c). Additionally, from Fig. 5.3, it can be seen that the host CBP emission intensity with the incorporation of the harvesting dopant is much reduced, suggesting more excitons generated are being utilized in the system. Without such green harvesting dopant, the η_{EQE} of the greenish-yellow dopant could only reach ~15%, whereas after the incorporation of the harvesting dopant, the efficiency was enhanced to over 21% (see Fig. 5.3a). This suggests a considerable amount of excitons leaked through the greenish-yellow EML, but were recaptured by the green dopants and delivered via energy transfer back to the greenish-yellow dopants. Note typically, this is a short-range (~1–2 nm) Dexter process; however, due to spin–orbit coupling where the singlet and triplet states are mixed, a long-range (~10 nm) Förster-type transfer process becomes efficient.[53,54] This demonstrate a simple way to enhance the performance of monochromatic organic light-emitting diodes (OLEDs).

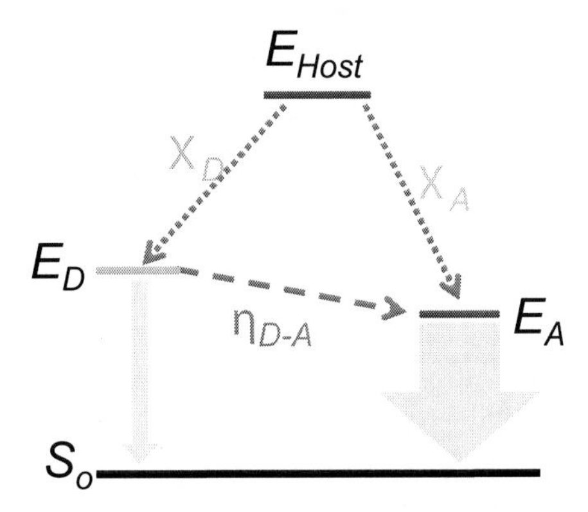

Figure 5.1 A simplified illustration of the various energy transfer processes (interzone or intrazone) taking place between two dopants in a common host. Dashed and dotted arrows represent nonradiative energy transitions and solid arrows represent radiative energy transitions. E_{Host}, E_A, E_D, and S_0 represent the energy levels of the host, acceptor, donor, and ground state, respectively. χ_A, χ_D, and $\eta_{D\text{-}A}$ are as defined for Eq. 5.1 in the text.

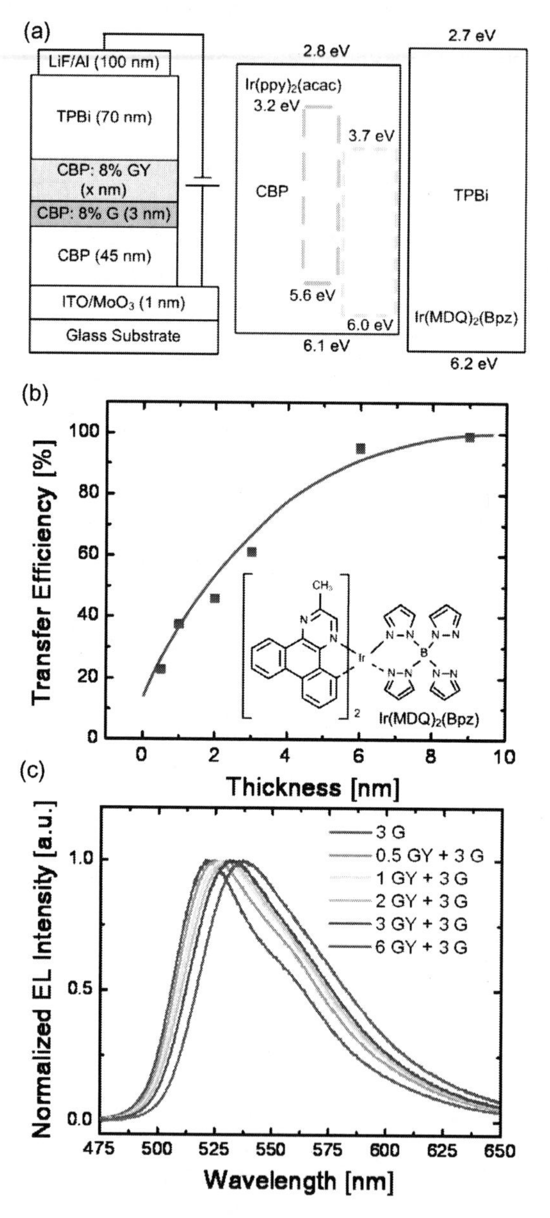

Figure 5.2 (a) Device structure utilizing interzone energy transfer for a greenish-yellow OLED. (b) Energy transfer efficiency with thickness of the greenish-yellow EML with corresponding EL spectra shown in (c). Reproduced with permission from Ref. [28]. Copyright 2013, Wiley-VCH.

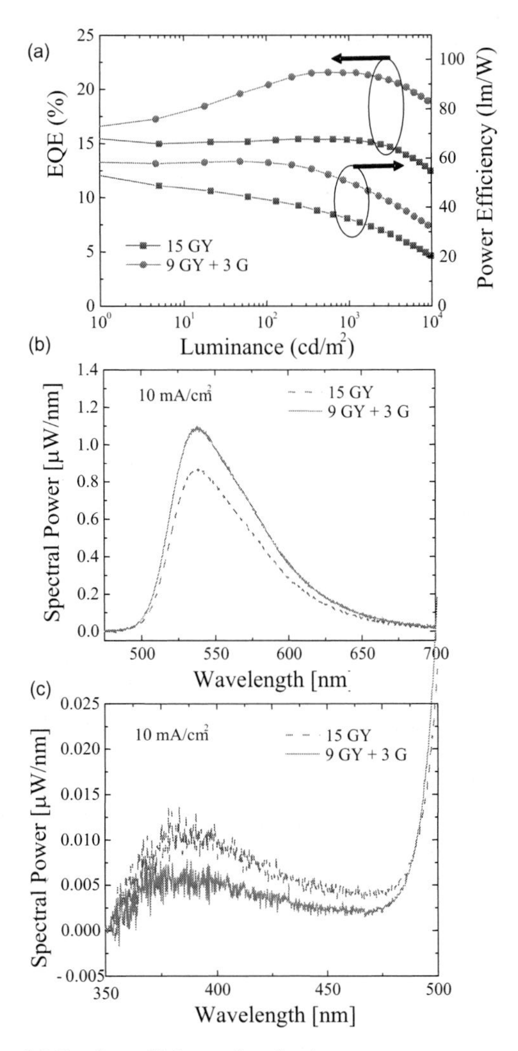

Figure 5.3 (a) Device efficiency for the interzone energy transfer–based device. Device EL spectrum (b) and host CBP emission (c) with and without the exciton-harvesting layer insertion. Reproduced with permission from Ref. [28]. Copyright 2013, Wiley-VCH.

Another closely related device architecture is shown in Fig. 5.4a, where the harvesting dopant is incorporated in the same spatial host region as the luminescent dopant. In this case, the spatial separation of the harvesting and luminescent dopant is even shorter, which facilitates the energy transfer process. This is called intrazone energy transfer,[52] which has the advantage of easier applicability to multiple emissive color OLEDs,[51] such as white OLEDs, as will be demonstrated in Chapter 8. Similar to the interzone case, the inclusion of the green harvesting dopant allowed both emission intensity enhancement and a remarkable increase in η_{EQE} of the luminescent (red) dopant at high brightness. At the same time, a small green emission from the harvesting dopant is also present which signifies saturation of triplet states of the red dopant (see inset of Fig. 5.4b).

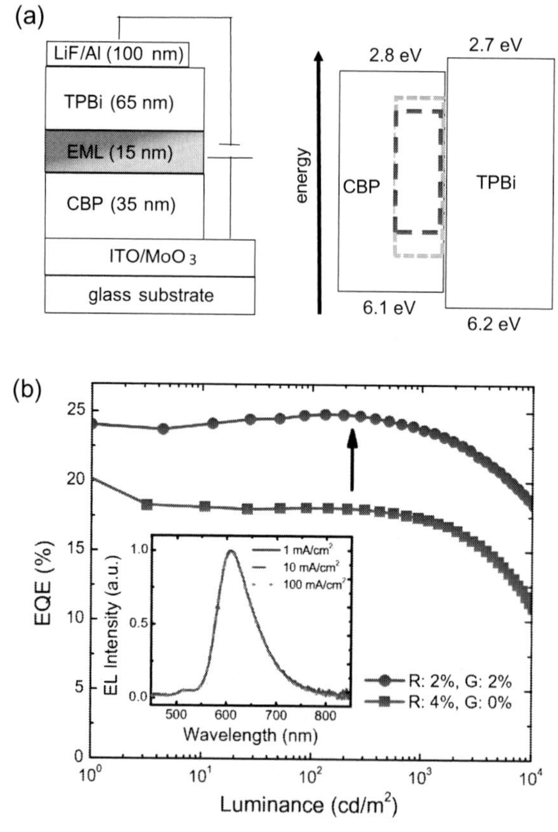

Figure 5.4 Device configuration (a), efficiency improvement (b), and EL spectra of a red phosphorescent OLED under varying current densities based on intrazone energy transfer. The EML is consisted of a green harvesting dopant, $Ir(ppy)_2(acac)$, and a red luminescent dopant, $Ir(MDQ)_2(acac)$. Reproduced with permission from Ref. [52]. Copyright 2012, Wiley-VCH.

5.2 Exciton Harvesting via TADF

As mentioned previously in Chapter 2, thermally activated delayed fluorescence (TADF) dopants can also harvest both singlet and triplet excitons similar to that of phosphorescent dopants. This type of dopant emerged only recently since it has long been a challenge to separate highest occupied molecular orbital (HOMO) and lowest unoccupied molecular orbital (LUMO) spatially within a single organic molecule in order to minimize orbital overlap and the exchange energy, while expecting a high fluorescent emissive yield. This feat was demonstrated to be possible by Adachi's group in 2012 with judicious molecular design.[3,23] Figure 5.5 shows the chemical structure of several TADF dopants that covers the entire visible range.

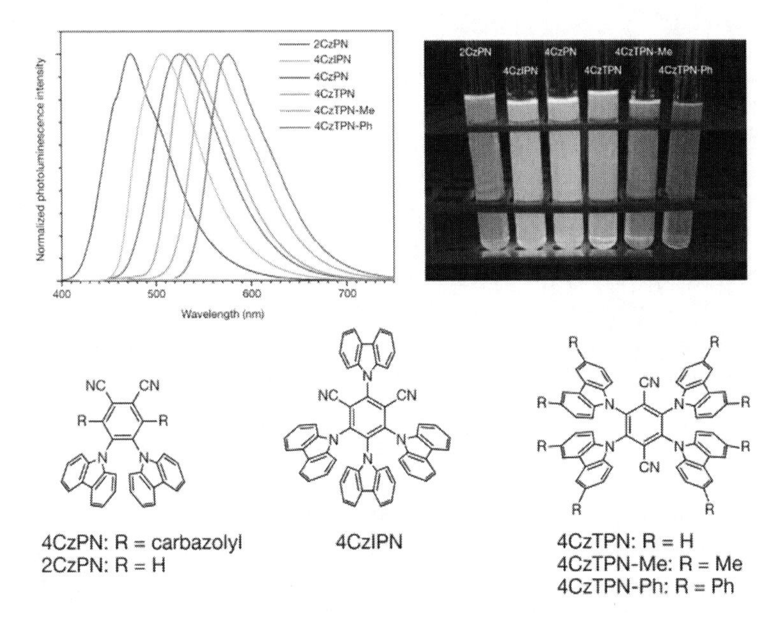

Figure 5.5 PL spectra, chemical structure, and photos of TADF dopants covering the entire visible range. Reproduced with permission from Ref. [3]. Copyright 2012, Nature Publishing Group.

To verify the presence of RISC in these molecules, transient photoluminescence (PL) studies were carried out, as shown in Fig. 5.6. Here, the prompt component increases very slightly as the temperature decreases, indicating the suppression of nonradiative decay from the S_1 state, which is typical of fluorescence emission. Conversely, a delayed component is also observed that decreases monotonically as the temperature decreases, which signifies the RISC process becoming the rate-limiting step, characteristic of TADF emitters. At room temperature (300 K), a high ϕ value, of ~83%, was observed. Such efficient TADF process is only possible due to a small singlet-triplet energy difference, ΔE_{ST}, which can be approximated by the activation energy of the RISC rate constant (k_{RISC}) from $\exp(2\Delta E_{ST}/k_B T)$, where k_B is the Boltzmann constant and T is temperature. This RISC rate constant can be estimated from experimentally determined rate constants as well as the ϕ values of the prompt and delayed components at each temperature from[3]

$$k_{RISC} = \frac{k_p k_d}{k_{ISC}} \frac{\phi_d}{\phi_p}, \tag{5.2}$$

where k_p and k_d are the rate constants of the prompt and delayed fluorescence components, respectively; k_{ISC} is the ISC rate constant from lowest singlet states, S_1, to the lowest triplet states, T_1; and ϕ_p and ϕ_d are the PL quantum yields of the prompt and delayed components, respectively. In Fig. 5.6d, the values of k_{RISC} were acquired from Eq. 5.2, assuming that k_{ISC} was independent of temperature, and plotted against $1/T$ for T = 200–300 K. From the Arrhenius plot (Fig. 5.6d), an activation energy of 80 meV can be obtained. Therefore, k_{RISC} would indeed be suppressed at low temperatures. This agrees well with the hypothesis that thermal energy is a key driving force for the RISC process.

It was demonstrated by Adachi's group that a green OLED using TADF dopants could achieve a remarkable η_{EQE} of over 20%, which is comparable to that of phosphorescent OLEDs. Through further judicious molecule design, it was even shown that blue OLEDs based on TADF dopants could achieve η_{EQE} of over 19.5% at high brightness,[23] which represents an enormous potential to replace phosphorescent blue dopants that remain to be the weakest link in phosphorescent dopant technology.

Recently, by taking advantage of exciton harvesting capability of TADF dopants, Nakanotani et al. have realized TADF assisted fluorescent OLEDs with remarkable performance.[55] Figure 5.7 shows the assistant TADF dopants and the illustration of the working mechanism. Here, once again, intrazone energy transfer is taking place between TADF and fluorescent dopants that are codeposited in the same host. The energy transfer process is a Förster type between TADF dopants' singlet states to the fluorescent dopants' singlet states. All triplet states remaining in the system are undesirable as they are nonemissive.

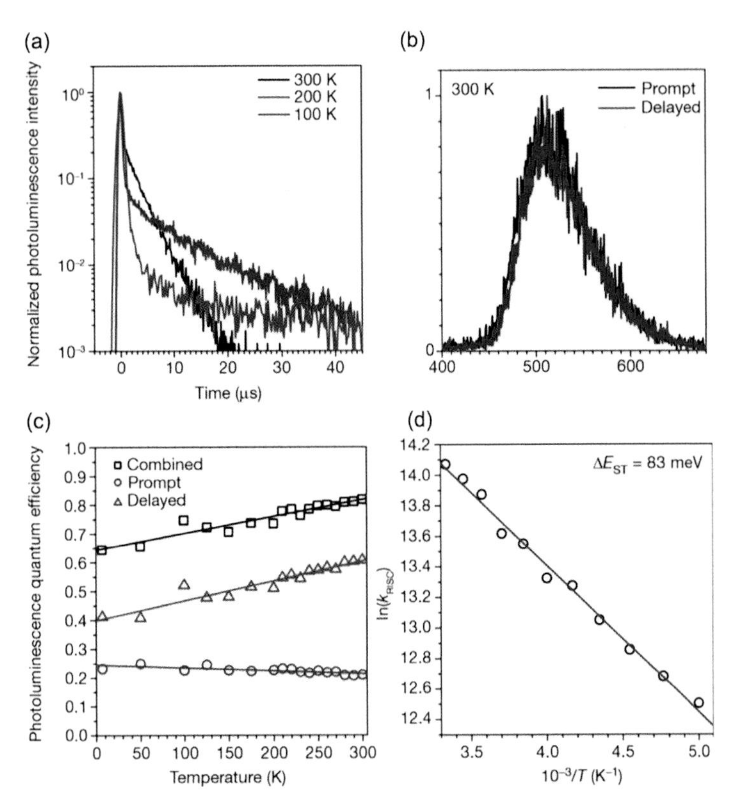

Figure 5.6 (a) PL decay curves of a 5 wt.% 4CzIPN:CBP film at 300 K (black), 200 K (red), and 100 K (blue). The excitation wavelength of the films was 337 nm. (b) PL spectrum resolved into prompt and delayed components. (c) Temperature dependence of PL quantum efficiencies for combined (prompt plus delayed; black squares), prompt (red circles), and delayed (blue triangles) components of 4CzIPN emission for a 5 wt.% 4CzIPN:CBP film. (d) Arrhenius plot of the RISC rate from the triplet state to the singlet state of 4CzIPN. The straight line is a least-squares regression used to determine the activation energy. Reproduced with permission from Ref. [3]. Copyright 2012, Nature Publishing Group.

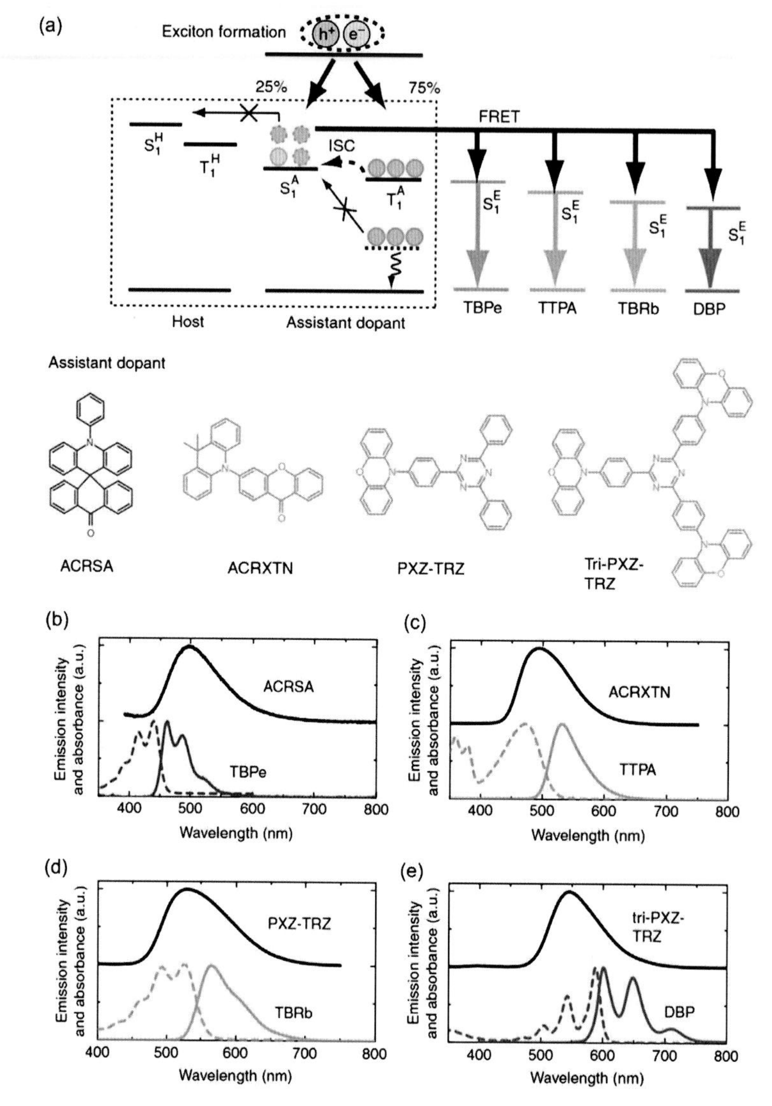

Figure 5.7 (a) Schematic illustration of the energy transfer mechanism in the emitter dopant:assistant dopant:host matrix under electrical excitation, as well as the assistant TADF dopant chemical structures. (b–e) Fluorescence spectra of assistant dopant:host codeposited film (upper) and absorption (dashed line) and fluorescence (solid line) spectra of emitter dopant in solution (bottom). Reproduced with permission from Ref. [55]. Copyright 2014, Nature Publishing Group.

Figure 5.8 shows a dramatic performance improvement with the use of TADF-assisting dopants for all four monochromatic OLEDs. Notice the use of four different assistant dopants for each color of fluorescent dopant. This is because energy transfer requires resonant energy level matching between the donor and acceptor molecules, analogous to molecular electron transfer process described in Marcus theory. For example, if a high-energy blue assistant TADF dopant is used to enhance the red fluorescent dopant that requires only a small energy input to excite its singlet states, a considerable waste in energy would be present in the system that

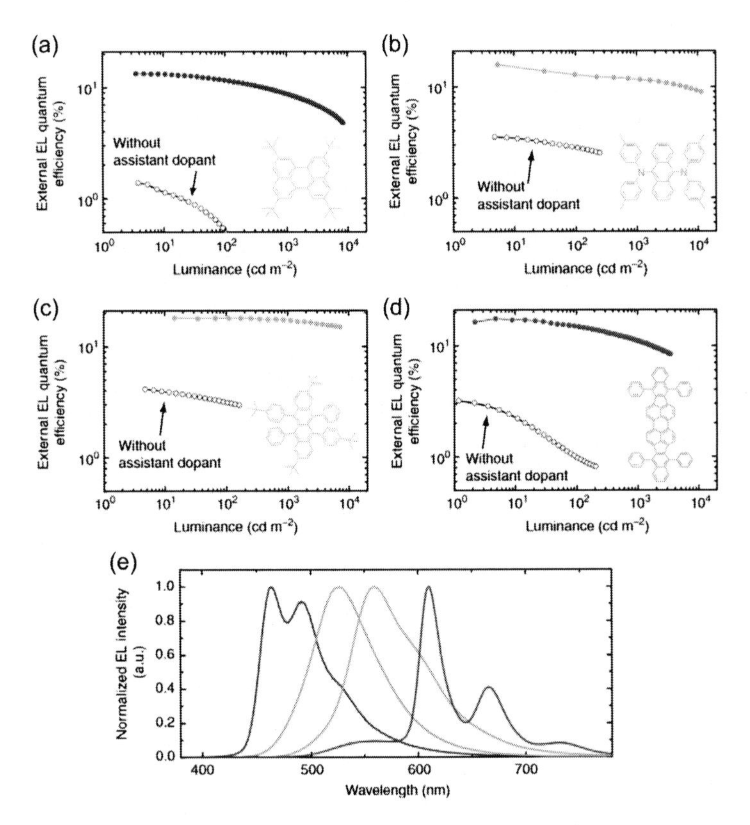

Figure 5.8 (a–d) EL efficiency as a function of luminance for the blue, green, yellow, and red OLEDs. Inset: Chemical structures of fluorescent dopants used in this study. (e) EL spectra of the devices at a luminance of 100 cd/m^2. Reproduced with permission from Ref. [55]. Copyright 2014, Nature Publishing Group.

generates heat from lattice vibrations or triggers various undesired quenching mechanisms. This fact can also be seen from the location of absorption peak of the fluorescent dopants in Fig. 5.7b–e, which is typically within ~50 nm from its emission peak. Conversely, if a red TADF assistant dopant is used to improve the blue fluorescent dopant, the assistant dopant would not have sufficient energy to promote the blue fluorescent dopants into excited states (lack of donor–acceptor spectral overlap).

As shown, the performance of TADF-based OLED is not far behind that of state-of-the-art phosphorescence-based OLEDs. In the case of TADF dopants as exciton harvesting or assistant dopants, the concentration employed can be as high as 50%, whereas the fluorescent dopant was only 1% with respect to the host.[55] It is then conceivable to simply utilize TADF as hosts directly. This also aligns with the fact that TADF dopants are low cost and efficient at harvesting excitons generated in the device. Zhang et al. have demonstrated OLED using TADF as the host to achieve high efficacy of 44.1 lm/W corresponding to an η_{EQE} of 11.7% for an orange fluorescent OLED.[24] A schematic illustration of the working mechanism is shown in Fig. 5.9. Once again the triplet states are not desirable and the Dexter triplet-to-triplet host–guest energy transfer is minimized by limiting the amount of triplets via RISC to singlets in the TADF host. On the basis of these studies, it can be anticipated that TADF hosts and dopants will be employed concurrently in an OLED to further eliminate problematic triplet states in such a cost-effective all-fluorescence system.

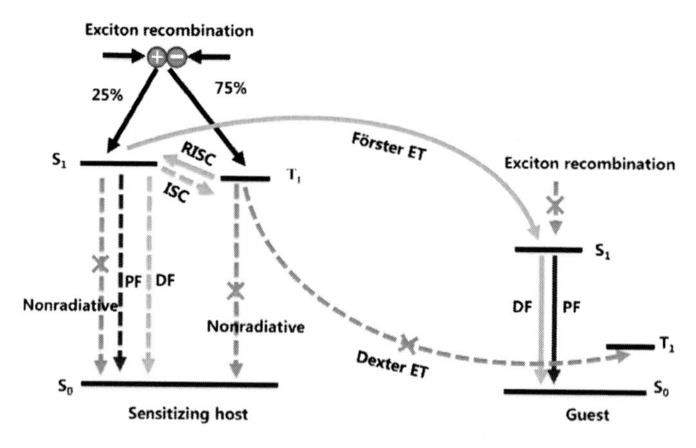

Figure 5.9 Working principles of a TADF host. Reproduced with permission from Ref. [24]. Copyright 2014, Wiley-VCH.

It is worth noting that phosphorescent dopants have also recently been used as hosts to produce high-efficiency orange OLEDs with low turn on voltages.[56] However, this approach is not economically feasible and typical high-energy (blue or UV emission) phosphors are prone to severe molecular instability and hence only applicable to long-wavelength emissions.

5.3 Exciton Harvesting via Exciplex-Forming Cohosts

In a standard OLED, the electrical carrier injection into the host material occurs on its HOMO via hole injection and LUMO by electron injection. Hence, the host material with a smaller energy gap (E_g), which approximately corresponds to the excited singlet energy (E_S), can reduce the driving voltage of OLEDs. Furthermore, as discussed previously, for the confinement of the exciton on a phosphorescent emitter, a host material with high triplet energy (E_T) is desirable (over 2.75 eV for blue). In essence, to reduce the driving voltage, hence increasing the power efficiency of phosphorescent OLEDs, the singlet energy, E_S, has to be as low as possible, while retaining a high E_T energy.[57] In other words, a small energy difference between E_S and E_T energies (the exchange energy, ΔE_{ST}) and high E_T energy are simultaneously desired to yield an effective host material.[57] In this regard, the use of an intermolecular interaction between electron donor and electron acceptor molecules, that is, exciplex formation,[58] is an attractive approach. It is interesting to note that for a long time the exciplex formation is considered to be a source of efficiency loss in OLEDs.[59] Recently, it was found that it could in fact be exploited to construct an ideal host material.[57] The exciplex has a small exchange energy due to a small orbital overlap between the HOMO and the LUMO, which are formed by two different host molecules mixed together to construct a single medium. An example of an effective exciplex system is shown in Fig. 5.10.[60–63] This exciplex-forming medium (TCTA:B3PyMBM) is essentially a third host material that has lower singlet and triplet energies than either hosts as indicated by a clear red shift of the PL emission compared to either hosts. In addition, it can be seen that this red shifted spectra is nearly identical both at room temperature (300 K) and at low temperature (35 K), suggesting minimal difference in singlet and triplet energy levels (very small exchange energy).

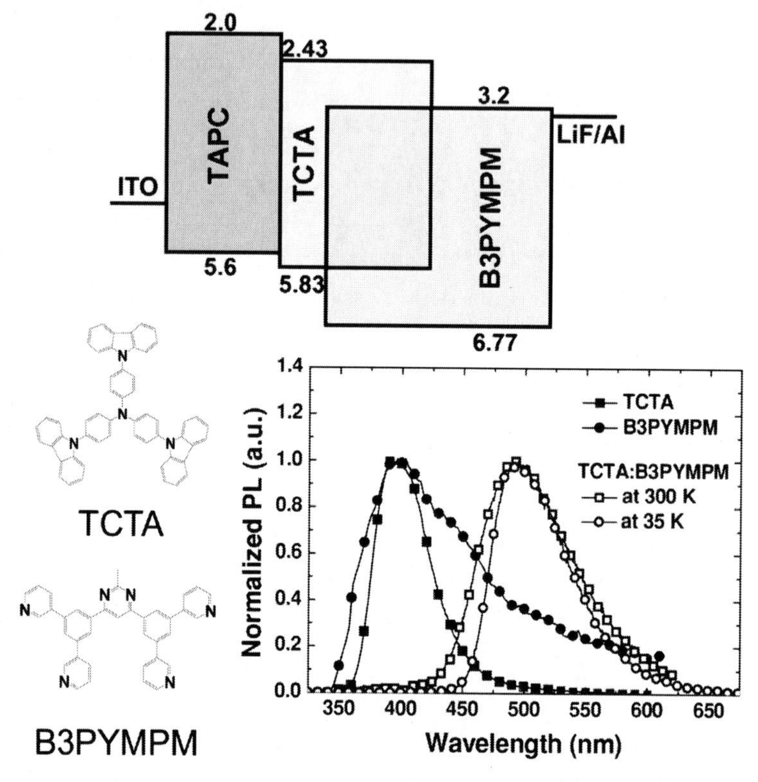

Figure 5.10 Device structure of a standard exciplex-forming cohost system featuring TCTA doped with B3PYMPM, where 1,1-bis-(4-bis(4-methyl-phenyl)-amino-phenyl)-cyclohexan (TAPC) chosen as the HTL. Reproduced with permission from Ref. [63]. Copyright 2013, American Institute of Physics. The chemical structure of the cohost materials and their temperature-dependent PL spectra are also shown. Reproduced with permission from Ref. [62]. Copyright 2013, Wiley-VCH.

One advantage of having small exchange energy is that very little thermal energy is needed to reverse intersystem cross the triplets generated in the material into singlets similar to that of a TADF material.[58,64] Second, by having mostly singlets in the host, the host–dopant energy transfer becomes dominated by a long-range Förster process, which suggests more efficient energy transfer (no need for donor–acceptor molecular orbital wavefunction overlap between host and dopant), and hence a lower dopant concentration compared to that used in phosphorescent system is needed. An energy-level diagram of an exciplex-based OLED is shown in Fig. 5.11.[65] Here, region II can be a mixture of hole-transporting (*p*-type) and electron-transporting (*n*-type) host materials or the interface between the two opposite monopolar host materials. From transient PL studies of a typical exciplex-forming cohost film shown in Fig. 5.12, a delayed component is observed especially at low temperature when vibrational losses are minimized.[58] This represents the contribution from RISC of the triplet states in the exciplex-forming cohost film that is facilitated by the small energy difference between singlet and triplet states.

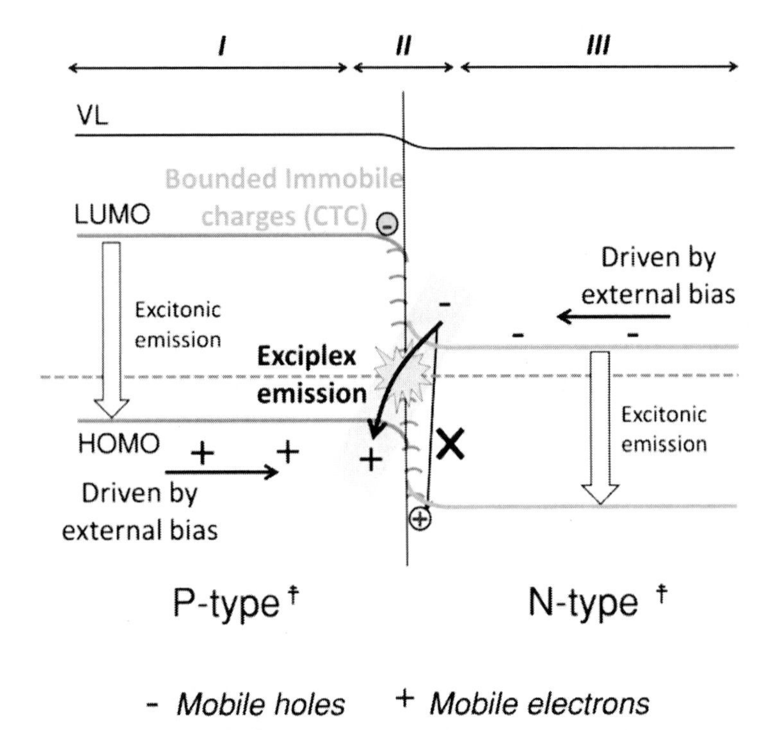

Figure 5.11 Energy-level diagram of an exciplex-forming *p-n* organic heterojunction. Reproduced with permission from Ref. [65]. Copyright 2014, Wiley-VCH.

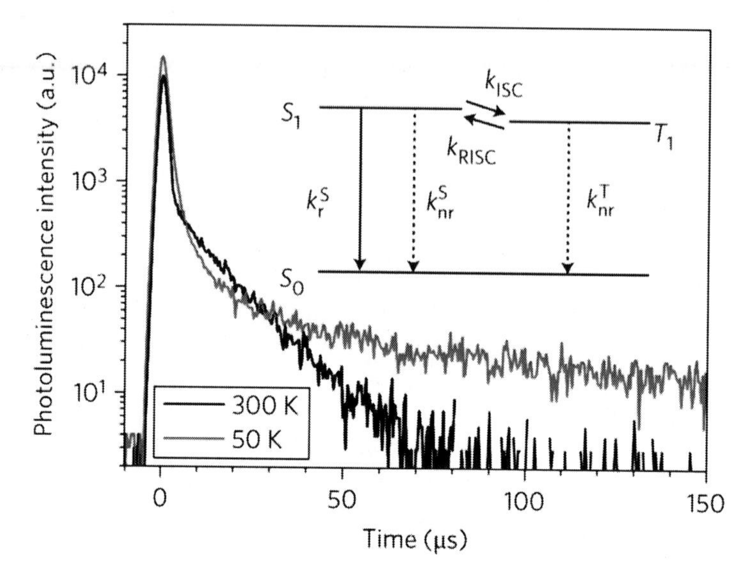

Figure 5.12 Temperature-dependent PL decay curves of an exciplex emission from a typical exciplex-forming cohost. Inset shows an energy diagram of the exciplex states. Reproduced with permission from Ref. [58]. Copyright 2012, Nature Publishing Group.

For the energy transfer mechanism, once again the spectrum overlap between the emission of the exciplex-forming host and the absorption of a guest is a key factor to obtain high PL quantum yield (η_{PL}). The energy transfer efficiency (ϕ_{ET}) of the singlet or the triplet exciplex to the guest molecules can be expressed as[57]

$$\phi_{ET} = \frac{k_{\text{exc-guest}}}{k_r + k_{nr} + k_{\text{exc-guest}}} = \frac{k_{\text{exc-guest}}}{\frac{1}{\tau} + k_{\text{exc-guest}}}, \tag{5.3}$$

where k_r is the radiative decay rate of the exciplex, k_{nr} is the nonradiative decay rate of the exciplex, $k_{\text{exc-guest}}$ is the energy transfer rate from the exciplex to the guest emitting molecule, and τ is the emission lifetime. From this equation, it is seen that a higher $k_{\text{exc-guest}}$ and a larger τ are required to achieve a high ϕ_{ET}. The typical fluorescent lifetime (τ_s) and phosphorescent lifetime (τ_t) of an exciplex emission are on the order of ~1.0 μs at 300 K and ~100 μs at 5 K, respectively. The fact that the singlet emission is significantly longer than that of a typical fluorescent molecule (~1 ns), suggests

RISC from the slow triplet states of the exciplex is efficient, which suggests a higher ϕ_{ET} is attainable from the exciplex according to Eq. 5.3. In other words, more (up-converted-from triplets) singlet states are available to be energy transferred to the dopants compared to a conventional host.

On the basis of the exciplex cohost system shown in Fig. 5.10, Kim's group has achieved an outstanding performance green phosphorescent OLED with a low turn-on voltage of 2.4 V, a very high η_{EQE} of 29.1% and a very high efficacy of 124 lm/W as shown in Fig. 5.13.[62] Even at an ultrahigh brightness of 10,000 cd/m^2, the η_{EQE} remains at 27.8%, which suggests a very small efficiency roll-off. This remains among the best-performance PHOLEDs to date in open literature. Shin et al. have even demonstrated record blue phosphorescent OLEDs of 29.5% η_{EQE}, which is close to the maximum theoretical limit using such exciplex-forming cohost architecture.[60] An additional advantage of this design is the fact that no extra layer of new material for carrier blocking or exciton confinement purposes is needed, which simplifies the fabrication process.

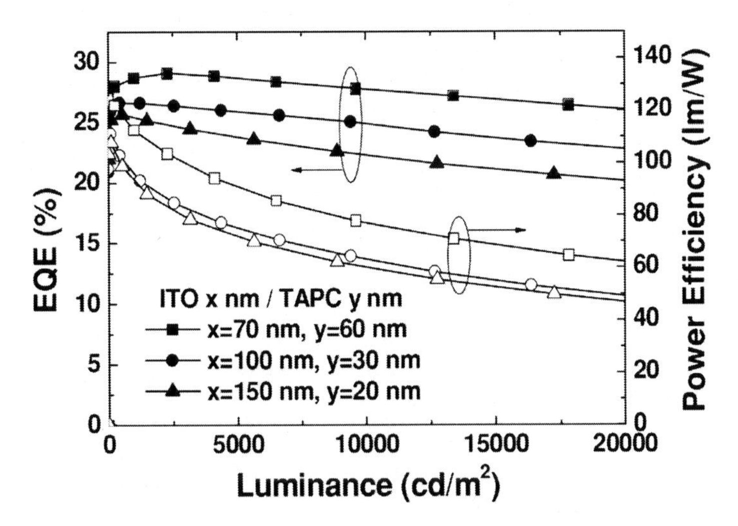

Figure 5.13 Device efficiency from an exciplex-forming cohost system shown in Fig. 5.10 using Ir(ppy)$_2$(acac) as the green phosphorescent dopant. These efficiencies are the highest reported to date for green phosphorescent OLEDs on ITO/glass substrates in open literature. Reproduced with permission from Ref. [62]. Copyright 2013, Wiley-VCH.

Interestingly, Sun et al. have also demonstrated the use of exciplex-forming cohost system with a TADF dopant, as shown in Fig. 5.14. Here, the exciplex-forming cohost is formed by codepositing mCP[*N,N′*-dicarbazolyl-3,5-benzene] and B3PYMPM,

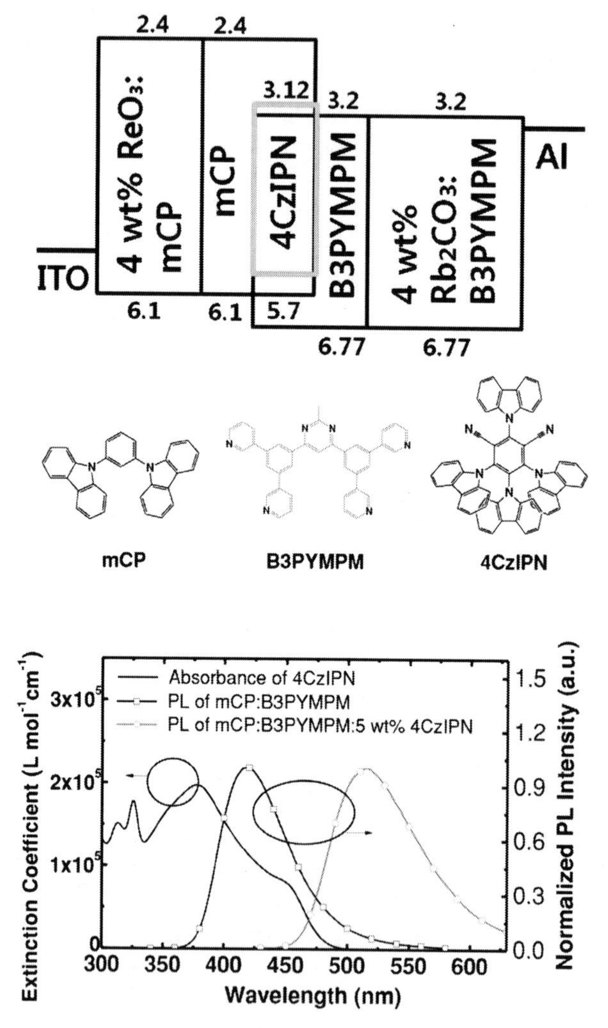

Figure 5.14 Exciplex-forming cohost system applied to a green TADF dopant. The chemical structures and the PL and absorption characteristics of the materials used are also shown. Reproduced with permission from Ref. [61]. Copyright 2014, Wiley-VCH.

with an emission that overlaps well with the absorption of the TADF dopant (4s,6s)-2,4,5,6-tetra(9H-carbazol-9-yl) isophthalonitrile (4CzIPN), as shown in Fig. 5.14, which ensures efficient host–dopant energy transfer. Notice that additional doping materials are incorporated in the transport layer of both hosts to further boost carrier injection and transport, which will be discussed in detail in the next chapter. Remarkable performance of 27.8% η_{EQE}, corresponding to an efficacy of 49 lm/W, was achieved at 1000 cd/m^2, which is the highest reported to date in TADF-based OLEDs. These results suggest that exciplex-forming cohost can conveniently provide an ideal host platform for both ultrahigh-efficiency TADF and phosphorescent OLEDs.

Nevertheless, it remains to be shown that these ultraefficient devices employing an exciplex-forming cohost can display a similarly long lifetime compared to conventional single-host devices using state-of-the-art phosphorescent and fluorescent dopants. This is because, unlike TADF materials or traditional fluorescent hosts that are formed mostly by covalent bonds, these exciplex-forming cohost molecules have to sustain high excitonic energy at the interface between two materials that are weakly bonded by van der walls forces. As with other advanced designs, intense effort is underway to achieve long lifetime OLEDs utilizing exciplex-forming cohosts.

Chapter 6

p-Type Intrinsic *n*-Type (*p*-*i*-*n*) OLEDs

Since an organic light-emitting diode (OLED) is a current-driven device with a lifetime that is inversely proportional to current density, simply achieving a high η_{EQE} at high current densities is insufficient for most commercial applications as it could nevertheless have an unacceptably short lifetime. It is therefore desirable to aim for high power efficiency with a low turn-on voltage such that the required brightness can be obtained at lower current densities, which would significantly extend the device lifetime.

In a standard OLED, the concentration of carriers is low, which leads to considerable Ohmic losses as well as a higher required field to drive the currents through the device. In some cases, as illustrated in the bottom part of Fig. 6.1b, a large field may be required across the device to drive the carriers, such that the operating voltage becomes a multiple of the photon energy emitted, which represents wasted electrical energy. In this regard, significant material design and synthesis work has taken place over the last two decades to develop highly conductive transport layers;[19] however, the conductivity is still not sufficient in most applications where low power consumption and long lifetime are critical. Fortunately, Blochwitz et al.[66] demonstrated that by doping the transport layers, not only the conductivity could be increased by several orders of magnitude, but charge injection could also be enhanced. This is because doped transport layers lead to very narrow space charge regions at the

Efficient Organic Light-Emitting Diodes (OLEDs)
Yi-Lu Chang

ISBN 978-981-4613-80-4 (Hardcover), 978-981-4613-81-1 (eBook)
www.panstanford.com

contacts, which allows for carriers to tunnel through, as shown on the top of Fig. 6.1b, such that Ohmic contacts are possible even with large energetic barriers across the contact. In fact, even for thick (~50–300 nm) doped layers, the voltage drop could become negligible and the length of the transport layers can then be used to optimize charge balance, define exciton formation zone location, and maximize optical out-coupling without the need for an additional injection layer. This type of doping technique is analogous to that for inorganic semiconductors such as Si or groups III–V.

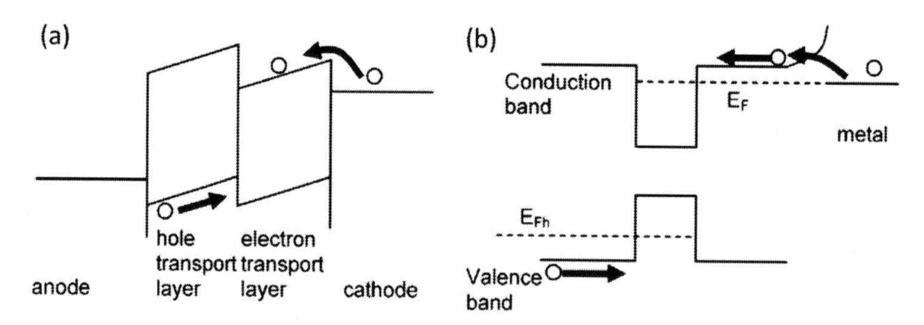

Figure 6.1 Energy band diagram of an OLED (a) and an inorganic LED (b). Reproduced with permission from Ref. [67]. Copyright 2009, American Chemical Society.

Figure 6.1 depicts energy band structures of an inorganic and an OLED device. In the inorganic light-emitting diode (LED) case (Fig. 6.1b), the emitter layer, characterized by a lower energy gap, is sandwiched between two highly *n*- and *p*-doped transport layers. Because of the high conductivity of these transport layers, Ohmic losses are negligible and the band edges are almost flat as there is nearly no voltage drop. The operating voltage of the device is then close to the photon energy of the light emitted.

In OLEDs, due to weak van der Waals binding of the molecules in solid form, there is no valence or conductive level as in inorganic solids. Electrons and holes move via a hopping mechanism between neighboring molecules.[14] Thus, charge carrier mobility and conductivity of the molecular layers are low. Therefore, one has to incorporate additional constituents that either donate extra electrons to the lowest unoccupied molecular orbital (LUMO) states (*n*-type doping) or remove electrons from the highest occupied molecular orbital (HOMO) states to generate additional holes (*p*-type doping), as illustrated in Fig. 6.2.[67] The energy levels of the dopants have to be carefully chosen to facilitate charge transfer to the matrix host material. The resulting organic layer should then display Ohmic behavior, $J = V/d$, and the voltage drop over the layer should be reduced and depends linearly on the layer thickness only. This is called *p-i-n* technology which has been extensively studied by Leo's group.[67,68]

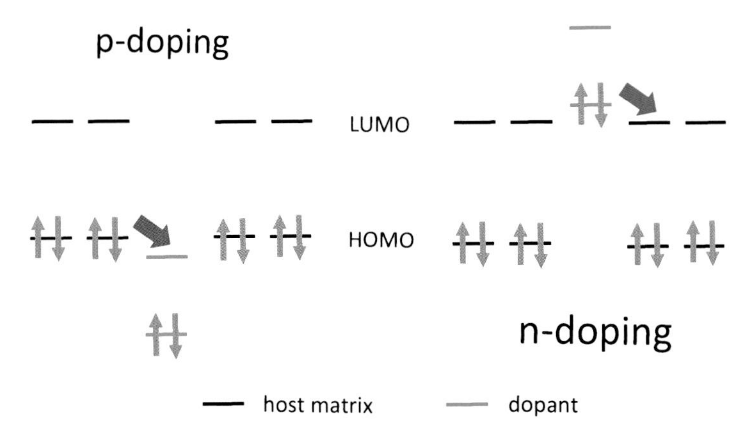

Figure 6.2 Charge transfer between a dopant and a host matrix. Given an appropriate location of the HOMO and LUMO energy levels, *p*-dopants will accept an electron, whereas *n*-dopants will donate an electron to the matrix.

A standard system for the realization of *p*-type doping is to apply ZnPc as the host matrix, and dope using 2,3,5,6-tetrafluoro-7,7,8,8-tetracyanoquinodimethane (F4-TCNQ), as presented by Pfeiffer et al.[69,70] The extremely large electron affinity of F4-TCNQ of ~5.2 eV allows electron acceptance from ZnPc, which exhibits a low ionization energy of 5.1 eV. Similarly, a slightly different hole-transporting material *N,N,N′,N′*-tetrakis(4-methoxyphenyl)-benzidine (MeO-TPD) can also be efficiently *p*-doped by F4-TCNQ where a doping ratio of 4 mol.% could lead to conductivities above 10^{-5} S/cm.[67]

Searching molecular *n*-dopants to reach such conductivity ranges is nontrivial.[71,72] The *n*-type dopants have to exhibit extremely high HOMO levels in order to donate electrons to the matrix LUMO level. Thus, the LUMO of the dopant is close to the vacuum level, which makes the dopant extremely unstable against oxygen. With increasing LUMO energy, the difficulty of finding suitable materials is only increased. Currently, a stable and highly conductive *n*-type doping system is commercially available from Novaled AG. These doped systems have already been adopted in active matrix organic light-emitting diode (AMOLED) displays worldwide. Conventionally, alkali metals such as LiF and Liq are used for an organic/cathode interface *n*-doping.[73] Another possibility is the doping of a matrix by lithium or cesium during the evaporation of the organic layer,[74] which could reach conductivities of above 10^{-5} S/cm. However, at higher operating temperatures, problems concerning the diffusion of atomic dopants arise, which is much more severe than using molecular dopants.[68]

Figure 6.3a shows an example of two *p-i-n* OLEDs stacked together to form a tandem structure with a thin charge generation unit connecting the two OLEDs.[75] The energy band diagram of such tandem structure is shown in Fig. 6.3b. Here, the key lies in utilizing highly doped layers to construct an effective charge generation unit that is able to allow high concentration of opposite carriers to tunnel through a narrow interface. After tunneling, the carriers to continue to travel to the next unit and contribute to exciton generation and radiative emission. Essentially, this is a means of recycling charge carriers, since each electron–hole pair produces two photons in such a tandem device.

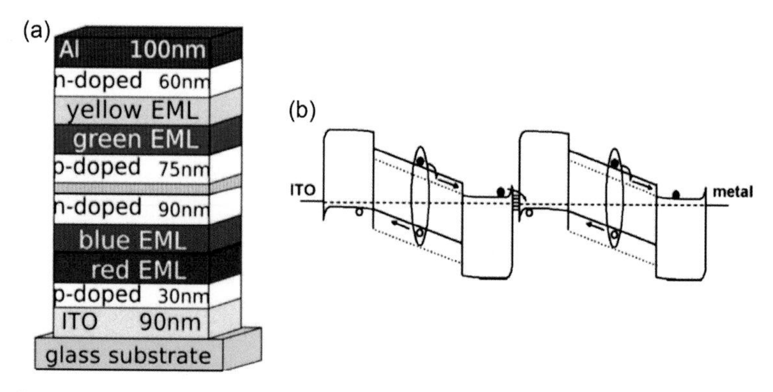

Figure 6.3 (a) A *p-i-n* white OLED with a double-unit, tandem structure. Reproduced with permission from Ref. [75]. Copyright 2013, Wiley-VCH. (b) The corresponding energy band diagram. Reproduced with permission from Ref. [67]. Copyright 2007, American Chemistry Society.

Chapter 7

Top-Emission OLEDs

In the currently prevalent active matrix organic light-emitting diodes (AMOLEDs), pixel driving circuits usually involve four or more backplane thin-film transistors (TFTs) combined with a capacitor. Such a large number of TFTs placed on the substrate competes with organic light-emitting diodes (OLEDs) in terms of space available on the chip, and inevitably reduces the aperture ratio of each pixel of a bottom-emission organic light-emitting diode (BEOLED), as shown in Fig. 7.1a,c. This reduced aperture ratio in turn results in the need to drive the OLED at a higher current density (which shortens device lifetime) in order to achieve the same level of luminance as a display having larger-aperture-ratio pixels. Fortunately, top-emission active matrix organic light-emitting diodes (TEOLEDs) can be employed to circumvent this issue since the TEOLED can be fabricated on top of any opaque substrate like silicon, thereby making it possible to place the TFTs below the reflective anode, as shown in Fig. 7.1b,d. Hence, using TEOLEDs, the aperture ratio can be significantly increased and the operating voltage required to achieve a desired luminance can be reduced, which further lead to longer device operational lifetime. Indeed this approach has been adopted with high success across current portable OLED display industries.

Efficient Organic Light-Emitting Diodes (OLEDs)
Yi-Lu Chang

ISBN 978-981-4613-80-4 (Hardcover), 978-981-4613-81-1 (eBook)
www.panstanford.com

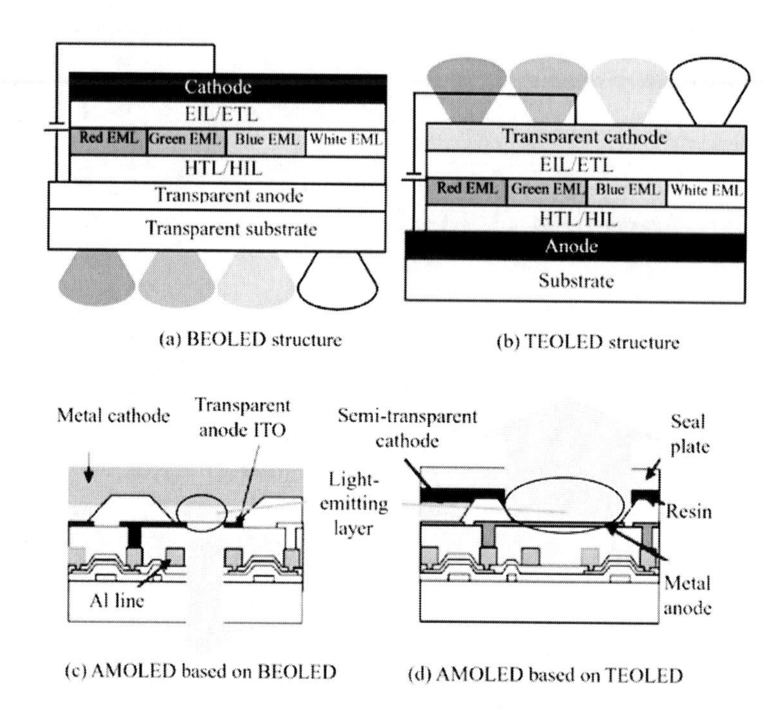

Figure 7.1 Schematic diagram of a BEOLED (a) and a TEOLED (b). Illustration of an AMOLED with low (c) and high (d) aperture ratios. Reproduced with permission from Ref. [76]. Copyright 2011, Wiley-VCH.

From Fig. 7.1a, it can be seen that for BEOLEDs, light is emitted through a transparent bottom contact, which is usually indium tin oxide (ITO). In TEOLEDs, however, a thick, highly reflective bottom contact is often used together with a semitransparent contact deposited on top of the organic layers. Using two highly reflective contacts, TEOLEDs inevitably show strong microcavity effects such as spectral narrowing and spectral shift of the emission peak, which translate into different emission spectra at different viewing angles.[77,78] This is compounded by the dispersion of light, where light with shorter wavelength bends at a sharper angle, leading to a blue-shift of the emission spectra at wider viewing angles. Nevertheless, such a strong microcavity effect can also be utilized to enhance emission intensity (and efficiency) in the forward viewing direction for a particular emissive color. However, in terms of performance, the TEOLED devices, for a long time, could not compete with their bottom-emitting counterparts. In fact, it was not until 2001 that similar performances for monochrome TEOLEDs were achieved by thermal evaporation of thin metal layers[79,80] together with the use of a dielectric capping layer (CL)[81,82] to extract the trapped modes of light out of the device. In 2010, an η_{EQE} of 29% for TEOLEDs,[83] which exceeds the efficiency of a similar BEOLED, was obtained for a phosphorescent red TEOLED by a judiciously optimized device stack.

Very recently, Schwab et al. have introduced a novel top transparent thin metal bilayer that minimizes microcavity effects, and further boosted the performance using a tandem *p-i-n* device structure, as shown in Fig. 7.2.[75] It was known for a long time that by forming a thin top metal contact with minimal absorption, yet still retaining a smooth surface for sufficient conduction, the microcavity effect would be greatly suppressed and less light would be trapped. However, this has not been possible for a long time since ultrathin metal films typically have poor conductivity. Here, it was demonstrated that an ultrathin (~2 nm) wetting layer of Au that was deposited before another thin (~7 nm) Ag layer to construct the top contact could yield an ideal transparent anode. From the atomic force microscopy images, it is seen that the bilayer contact was considerably thinner and smoother than a relatively thick Ag single layer (see Fig. 7.2b). In addition, such bilayer cathode exhibits transmittance across the visible range that is on par with that of the state-of-the-art ITO, as shown in Fig. 7.2b. This led to a device that is superior to traditional TEOLED with a single Ag top contact, as shown in Fig. 7.2c, which represents a breakthrough in TEOLED technology. Yet its performance is still slightly behind that of the BEOLED employing an ITO/glass contact.

To understand the microcavity effect in TEOLEDs, it is best to model such a structure as a Fabry–Perot resonator, where the anode and cathode are considered two parallel mirrors. In such structure, the out-coupled emission spectrum will display an approximated full width at half maximum (FWHM)[84] of

$$\mathrm{FWHM} = \frac{\lambda^2}{2L_{\mathrm{cav}}} \times \frac{1-\sqrt{R_\mathrm{T}R_\mathrm{B}}}{\pi\sqrt[4]{R_\mathrm{T}R_\mathrm{B}}} \tag{7.1}$$

Here, λ is the peak emission wavelength, $L_{cav} = nd$ (n is the refractive index and d is the cavity thickness) denotes the optical cavity thickness, and R_B and R_T represent the reflectivity of the bottom and top contacts, respectively. It is clear from Eq. 7.1 that

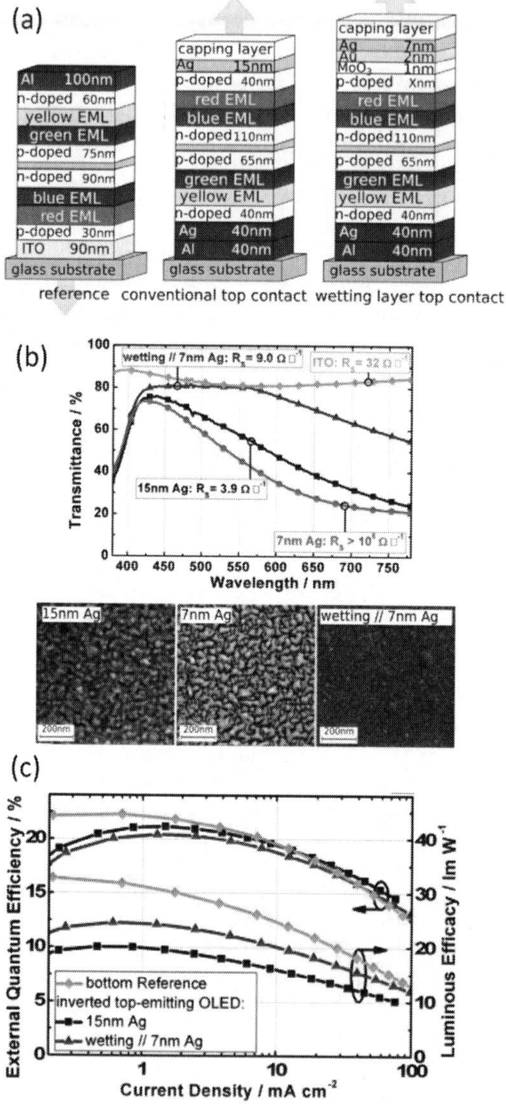

Figure 7.2 (a) Device structure of a BEOLED, a TEOLED with a conventional cathode, and a TEOLED with a novel thin bilayer cathode. (b) Transmittance of the ITO and various bilayer metal thicknesses and atomic force microscopy images of the thin bilayer metals. (c) Device performance of the three device structures in (a). Reproduced with permission from Ref. [75]. Copyright 2013, Wiley-VCH.

if the cavity length or the reflectivity of either contact is increased, the FWHM is reduced. For example, given a reflectivity of $R_T = 0.5$, $R_B = 0.9$, $n = 1.7$, and a cavity length d of 100 nm, one would obtain a FWHM of 240 nm (60 nm) for a peak emission wavelength of 800 nm (400 nm). It is conceivable that designing a white TEOLED with a broad emission across the entire visible range (380 nm to 780 nm) is a considerable challenge.

The intensity of device spectral emission $I(\lambda,\theta)$ arising from microcavity is dependent on the emission wavelength and the emission angle, θ, measured from normal (forward) direction, as described by[85]

$$I(\lambda,\theta)=I_o(\lambda)\frac{T_T[1+R_B+2\sqrt{R_B}\cos(-\varphi_B+\frac{4\pi nz\cos(\theta_{\mathrm{org,EML}})}{\lambda})]}{4\sqrt{R_T R_B}\sin^2(\frac{\Delta\varphi}{2})+(1-\sqrt{R_T R_B})^2}, \quad (7.2)$$

where T and R represent the transmittance and reflectivity of contacts, respectively, φ_B is the phase shift at the bottom contact, and z denotes the distance from the emitter to the highly reflecting bottom contact. The factor I_o represents the emission of the radiating molecules. After one period of light propagation, the total phase shift $\Delta\varphi$ can be found by[83]

$$\Delta\varphi = -\varphi_B - \varphi_T + \sum_i \frac{4\pi n_i d_i \cos(\theta_{\mathrm{org},i})}{\lambda} = 2\pi m. \quad (7.3)$$

Here, m is the mode index, d_i denotes the thicknesses, and n_i is the refractive indices of each organic layer inside the cavity. It is clear that the radiated intensity I is dependent on the emission angle θ, which is directly related by Snell's law to the propagation directions $\theta_{\mathrm{org},i}$ of the light within the organic layers, and also related to the angle $\theta_{\mathrm{org,EML}}$ in the emitting layer. On the basis of these equations, it can be understood that to achieve white light emission using three primary color emissions in one stack, one cannot simultaneously achieve the maximum emission intensity for each primary color with a single fixed resonant cavity structure.

If we simplify the calculation by considering only forward emission, Eq. 7.2 becomes[86]

$$I(\lambda)=I_o(\lambda)\frac{T_T\left[1+R_B+2\sqrt{R_B}\cos\left(\frac{4\pi nz}{\lambda}\right)\right]}{1+R_BR_T-2\sqrt{R_TR_B}\cos\left(\frac{4\pi L_{cav}}{\lambda}\right)}. \qquad (7.4)$$

It can be realized that to obtain constructive interference, the cavity length L_{cav} has to be chosen as multiples of $\lambda/2$, as shown in Fig. 7.3. This also means the emissive layer (EML) has to be positioned at an antinode to maximize light out-coupling in the forward direction for the chosen wavelength of emission.

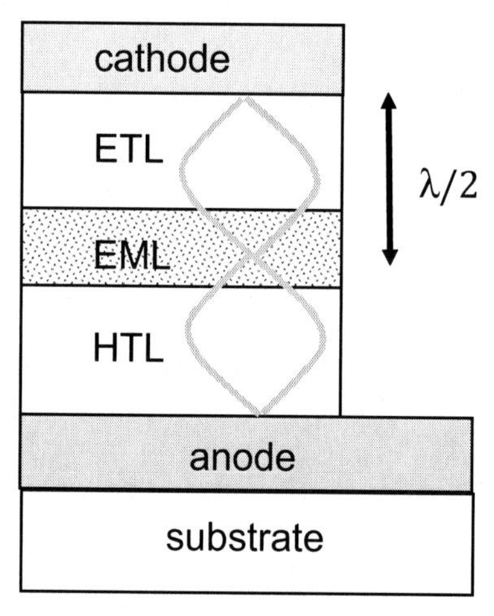

Figure 7.3 A schematic diagram of the ideal device length to maximize the out-coupling of light in the forward direction for a given wavelength, λ, according to Eq. 7.4.

By considering the Purcell effect, the emission enhancement factor G with respect to free space emission at a wavelength λ can be found by[84]

$$G=\frac{\tau_{cav}}{\tau}\times\frac{T_T\left[\left(1+\sqrt{R_B}\right)^2-4\sqrt{R_B}\cos^2\left(\frac{2\pi z}{\lambda}\right)\right]}{\left(1-\sqrt{R_TR_B}\right)^2+4\sqrt{R_TR_B}\sin^2\left(\frac{2\pi L_{cav}}{\lambda}\right)}. \qquad (7.5)$$

Here, τ_{cav} is the exciton lifetime in the cavity and τ denotes the lifetime in an infinite medium. From Eq. 7.2, it can be deduced that the spectral emission may display a blue shift, $\Delta\lambda_\theta$, of the peak wavelength upon increasing viewing angle θ (away from the normal, forward direction), which can be approximated by[13]

$$\Delta\lambda_\theta = \sum_\theta \left[\sum_i \frac{4\pi n_i d_i}{\lambda} \cos(\theta_{\mathrm{org},i} - 1) + \Delta\varphi_\mathrm{B} + \Delta\varphi_\mathrm{T} \right], \tag{7.6}$$

In addition to material selection for the TEOLED, the emission intensity and device efficiency also depend strongly on the cavity length, the position of the emitter molecules relative to the bottom contact, and the thickness of the top contact and capping layer. Note that the capping layer and the top contact thicknesses can be adjusted without altering the electrical properties of the TEOLED. In addition, if *p-i-n* doping technology is employed, it is also possible to tune the cavity length and the position of emitter molecules without significantly compromising the low driving voltage.[83] As will be discussed in Chapter 9, a number of external light extraction techniques such as the use of a capping layer embedded with light-scattering centers[87] or a lamination foil comprising of microlens arrays[88] are effective for simultaneously achieving high efficiencies and better angular stability of the emission colors.

It is worth noting that for low cost display driver circuits using amorphous Si TFT technology, having an *n*-type channel offers more stability and hence is preferred. This suggests in display applications, OLED structures should have an inverted structure with the cathode located on the substrate, followed by the ETL and so on (Fig. 7.2a). Thus, the last anode layer on top has to be consisted of either semitransparent metal or conductive oxides such as ITO for light out-coupling purposes. This has proven to be a challenge since sputtered ITO would damage the organic layers underneath. In addition, by simply depositing organic layers on top of the cathode and the electron injection layer, there is a lack of metal penetration into the organic layers, thereby leaving a sharp cathode–ETL interface such that electron injection is quite poor.[89] These facts led to most inverted OLED devices having a considerably higher driving voltage and lower power efficiency compared to their noninverted counterparts. In this regard, a number of approaches including the use of ZnS nanoparticles,[90] insertion of a pentacene interlayer,[91] or employment of a tris(8-hydroxyquinolinato) aluminum(Alq) + LiF + Al trilayer[89] have been shown to improve electron injection in inverted OLEDs. Alternatively, alkali metal-doped injection and transport layers,[92] as well as WO_3 buffer layers[93] insertions are effective ways to improve the device performance. Recently, thermal annealing has also been found to lower the driving voltages significantly.[94] In such annealing process, material diffusion and morphological changes are responsible for the enhancement in device performance.

Chapter 8

Efficient White OLEDs

Artificial lighting is an integral part of our lives that consumes nearly one-fifth of the planet's electricity usage.[95] Hence, by using more efficient devices that require less electricity, it is possible to dramatically reduce energy costs, thereby greatly cutting down global CO_2 production. In this regard, ultraefficient lighting[95] such as white organic light-emitting diodes (OLEDs) will play an important role. In the following, major types of OLED architectures for efficient white-light generation is discussed.

8.1 Single-Emissive-Layer White OLEDs

One straightforward approach to achieve white organic light-emitting diodes (OLEDs) is to codeposit three phosphorescent emitters[96,97] into a single host to form a single emissive layer (EML), as demonstrated by D'Andreade et al.,[96] shown in Fig. 8.1a. To contain the high triplet energy of the deep blue emitter, bis(2,4-difluorophenylpyridinato) tetrakis(1-pyrazolyl)borate iridium(III) (FIr6), and to prevent significant back energy transfer from the dopant to the host, an ultrawide-energy-gap host material, UGH2 [1,4-bis(triphenylsilyl)benzene], was used. The dopant concentrations in percentage weight with respect to the host for blue, green, and red dopants were 20%, 0.5%, and 2%, respectively. The reason for such

Efficient Organic Light-Emitting Diodes (OLEDs)
Yi-Lu Chang

ISBN 978-981-4613-80-4 (Hardcover), 978-981-4613-81-1 (eBook)
www.panstanford.com

high blue and low green and red doping concentrations is to limit the number emissive sites of the low-energy emitters available such that the energy transfer from the blue dopant to green and red dopants are reduced. In this way a balanced white electroluminescent (EL) spectrum can be obtained, as shown in Fig. 8.1b. At 10 mA/cm^2, a Commission Internationale de L'Eclairage (CIE) coordinates of (0.38, 0.45) and a CRI of 78 is demonstrated. However, the η_{ext} and efficacy of the device at 1000 cd/m^2 are reported to be only 6.7% and 11.1 lm/W, respectively. One reason for the low efficiencies is that while successful at preventing energy back transfer, this ultrawide-energy-gap host often exhibits poor mobility and stability, leading to higher charge carrier imbalance and a serious device efficiency

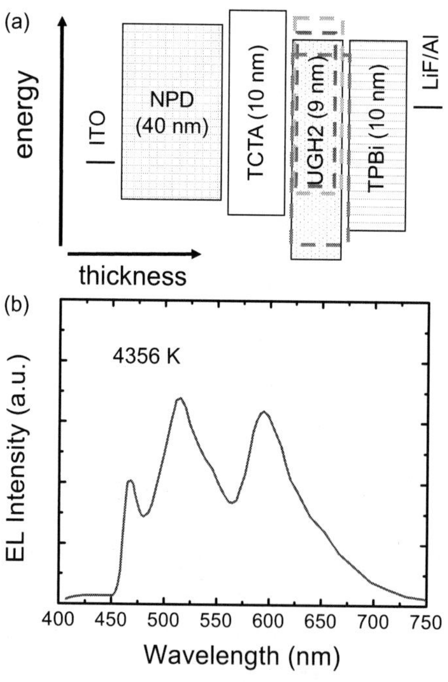

Figure 8.1 (a) Energy-level diagram and device structure of a single EML white OLED and (b) the device EL spectrum at 10 mA/cm^2. The blue, green, and red dopants used are FIr6, Ir(ppy)$_3$, and iridium(III) bis(2-phenylquinolyl-*N*,C$^{2'}$) acetylacetonate (PQIr), respectively. 4,4'-bis[*N*-(1-naphthyl)-*N*-phenyl-amino]-biphenyl (NPD) is used as the HTL. This figure is generated with permission from Ref. [96]. Copyright 2004, Wiley-VCH.

droop. Furthermore, such codoping of three emitters in one emission zone limits severely the degree of freedom in tuning the doping concentrations of each emitter due to the need for a balanced white emission. Such limitation hinders device efficiency optimization, thereby leading to a lower performance.

8.2 Hybrid White OLEDs

One promising approach to achieve white OLED is to make use of a fluorescent blue emitter together with phosphorescent green and red emitters to construct a three-color white device, or also referred to as a hybrid white OLED.[98-101] The advantage here is that for a practical warm-white illumination, only roughly 25% of the light needs to come from the blue portion, which is coincidentally also the fraction of singlet excitons that could be converted to light by a standard fluorescent dopant. This design concept in theory could provide nearly 100% internal quantum efficiency for the white device as a whole, and also should lead to a longer device operation lifetime due to absence of short-lived blue phosphorescent dopants. Figure 8.2a shows a device configuration using this design. To prevent energy transfer from the blue fluorophore to the red and green phosphors, one or more interlayers or spacers consisting of nondoped host CBP are inserted. The EL spectrum for the device at 10 mA/cm^2 is shown in Fig. 8.2b, with CIE coordinates of (0.39, 0.40) and a CRI of 85. In general, since the blue fluorescent dopant has a wider FWHM and a deeper blue peak (higher saturation) than standard blue phosphorescent dopants, the CRI of hybrid white OLED is typically quite decent (>85).[101] In this case, at a luminance level of 500 cd/m^2, the η_{EQE} and efficacy achieved are 18.4% and 23.8 lm/W, respectively. Here, the efficiency limiting factors in this design include the inefficient fluorescent dopants (η_{EQE} of ~2.6%) used and the undesirable emissions from the spacer regions where a substantial amount of generated excitons are used by the host CBP to emit inefficiently. Additionally, because of the intrinsic low efficiency of the blue fluorescent dopant, the white illumination could at best be a balanced or pure white with CIE coordinates of (0.33, 0.33) at low luminance levels adequate for displays. It can only offer a warm white illumination with a weak blue emission at higher luminance levels (~3000–5000 cd/m^2) required for solid-state lighting. Essentially, the hybrid white OLED design would not provide an efficient daylight white illumination characterized by a dominant blue emission (higher intensity than green and red) at a brightness required for lighting applications. Nevertheless, one critical advantage with the use of blue fluorescent dopant is that it requires hosts with a lower energy gap (singlet energy), just sufficient enough to excite the blue dopant. Conversely, if a phosphorescent blue is used, a host with a high triplet energy (and necessarily an even higher singlet energy or energy gap) would be required. By avoiding the use of such ultrawide-energy-gap organic molecules, the charge injection, operational lifetime, and molecular stability can be greatly improved.

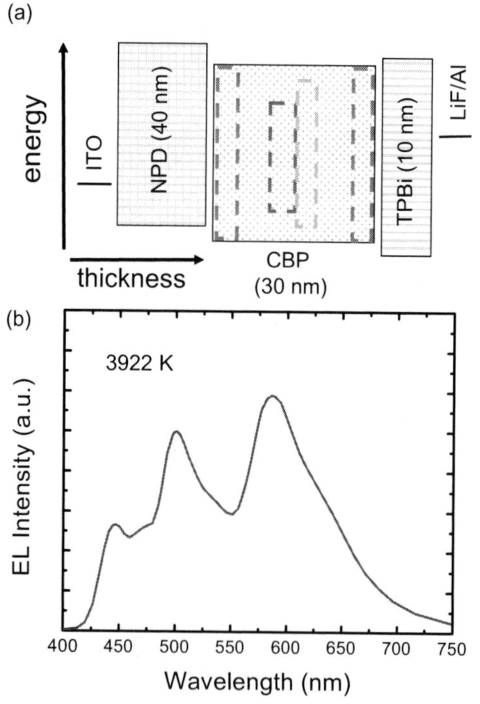

Figure 8.2 (a) Energy-level diagram and device structure of a hybrid white OLED and (b) the device EL spectrum at 10 mA/cm^2. The blue, green, and red dopants used are 4,4'-bis(9-ethyl-3-carbazovinylene)-1,1'-biphenyl (BCzVBi), Ir(ppy)$_3$, and PQIr, respectively. This figure is generated with permission from Ref. [99]. Copyright 2006, Nature Publishing Group.

8.3 Stacked White OLEDs

Another currently popular design has been demonstrated initially by Kanno et al.[102] called a stacked or tandem white OLED,[102–105] as shown in Fig. 8.3. The key difference is the insertion of a charge generation interlayer consisted of MoO_3 and a doped

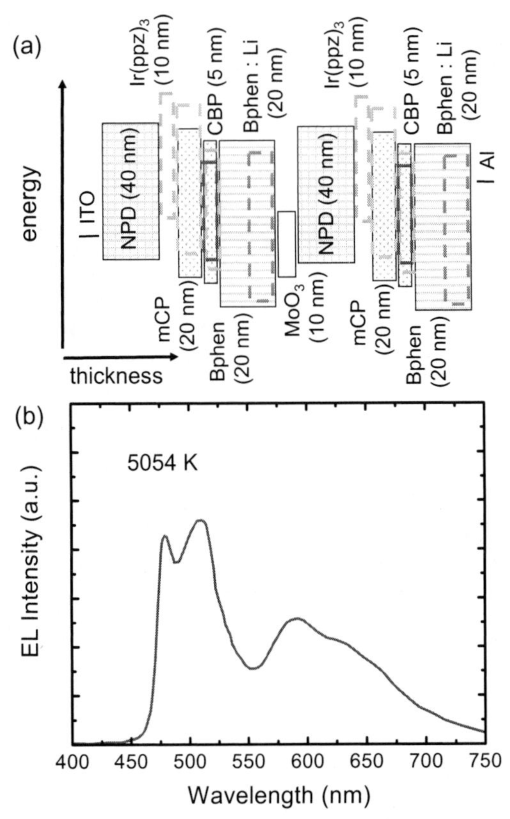

Figure 8.3 (a) Energy-level diagram and device structure of a tandem white OLED and (b) the device EL spectrum at 10 mA/cm^2 for a three-unit device. The blue, bluish-green, green, and red dopants used are fac-tris(1-(9,9-dimethyl-2-fluorenyl)pyrazolyl-*N*,C$^{2\prime}$) iridium(III) ($Ir(flz)_3$), $Ir(ppz)_3$[fac-tris(1-phenypyrazolyl-N,C$^{2\prime}$) iridium(III)], $Ir(ppy)_3$, and PQIr, respectively. mCP is a wide-energy-gap host used to contain the high triplet energy of the blue dopant. This figure is generated with permission from Ref. [102]. Copyright 2006, Wiley-VCH.

electron transport layer (ETL), namely, Bphen [4,7-diphenyl-1,10-phenanthroline]:Li, between two (shown in Fig. 8.3a), three, (or more) repeating white OLED units. In this design, the luminance at a fixed current density linearly increases with the number of repeating white OLED units, which consequentially results in a lower driving current required to obtain a comparably high brightness of a single unit, thereby significantly reducing numerous high-current-induced degradation processes and enhancing the overall device operation lifetime. The EL spectrum of a three units stacked white OLED device at 10 mA/cm^2 is shown in Fig. 8.3b, where CIE coordinates of (0.35, 0.44) and a CRI of 66 are demonstrated. The η_{EQE} and efficacy at 1000 cd/m^2 are ~27% and ~11 lm/W, respectively. It is worth noting that since each white OLED unit is essentially connected in series, a high driving voltage that is representative of the number of repeating units is observed. To compensate for this high voltage operation, note that a smaller current density is needed to achieve high brightness owing to the charge-recycling property provided by the charge generation layers in the stack in order to maintain a high power efficiency. In addition, efficient doping strategies are necessary to maintain a high charge transport throughout the thick layers to further lower the operating voltage. One caveat of the design is the cumbersome fabrication procedure that leads to a significant degree of difficulty in minimizing microcavity effects formed in the device, thereby severely hindering the out-coupling fraction of light. Recently, Lee et al.[105] have achieved outstanding-performance white OLEDs with a maximum η_{EQE} of 54.3%, corresponding to a efficacy of 63 lm/W with the use of effective doping, and charge generation units. Indeed this approach has become the preferred choice to produce white OLEDs in industry.

8.4 Multiple-Emissive-Layer White OLEDs

8.4.1 Separate Emissive Layers

From the industrial production point of view, it is more ideal to employ simpler device designs having less total host and transport layers, while preserving high efficiency and lifetime.[43,51,106–108] One approach demonstrated by Reineke et al.[106] is shown in Fig. 8.4a, where three separate, phosphor-doped EMLs are used to facilitate tuning of a desirable spectrum by both doping concentration and thickness of the EML within a single OLED unit. Thin (~2 nm) spacers or nondoped host layers (4,4′,4″-tris(carbazol-9-yl)triphenylamine [TCTA] and 2,2′,2″-(1,3,5-benzinetriyl)-tris(1-phenyl-1-H-benzimidazole) [TPBi] in this case) are once again introduced to prevent Förster and/or Dexter energy transfer from the blue phosphorescent dopant to the nearby red and green phosphorescent dopants, such that the blue emission could remain relatively strong. The EL spectrum at 1000 cd/m^2 is presented in Fig. 8.4b, where ideal warm-white CIE coordinates of (0.44, 0.46) and a decent

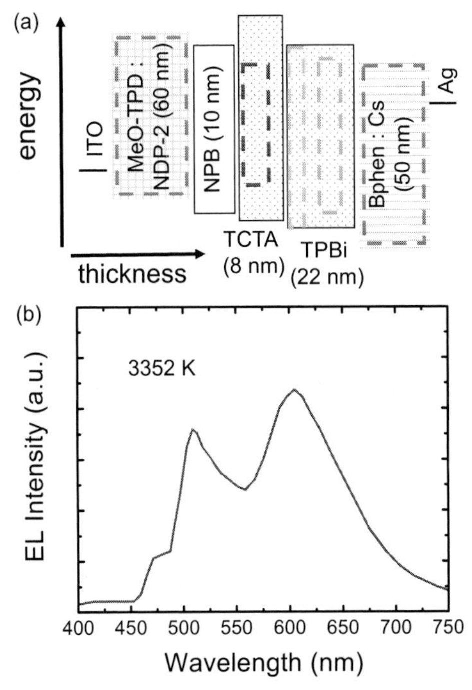

Figure 8.4 (a) Energy-level diagram and device structure of a separate emissive layer white OLED and (b) the device EL spectrum at 10 mA/cm^2. The blue, green, and red dopants used are iridium bis-(4,6,-difluorophenyl-pyridinato-*N*,C$^{2'}$)-picolinate (FIrpic), Ir(ppy)$_3$, and PQIr, respectively. NPB is used as the electron-blocking layer. MeO doped with NDP-2 is used as the hole transport layer, and Bphen doped with Cs is employed as the electron transport layer. This figure is generated with permission from Ref. [106]. Copyright 2009, Nature Publishing Group.

CRI of 80 are achieved. The corresponding η_{EQE} and efficacy are 14.4% and 33 lm/W, respectively. Owing to the incorporation of doped electron and hole transport layers that improve charge transport mobility, and the relatively thin EMLs, the power efficiencies are quite decent. However, the low η_{EQE} could be attributed once again to the exciton utilization by the inefficient host emission in the nondoped interlayers, as well as potential quenching effects induced by the dopants in the transport layers. It is worth noting that while high efficiency can be obtained using such all-phosphorescent emitter approach, the device lifetime is still limited by the short-lived blue phosphorescent dopant.

8.4.2 Cascade Emissive Layers

To preserve the device simplicity and improve the performance of white OLEDs, the author has recently developed a simplified, cascade EML structure using a total of only two host and transport materials together, as shown in Fig. 8.5a.[51] Here, four separate phosphor-doped EMLs are utilized to allow freedom in tuning the emission strength of each emitter using doping thickness and concentration. More importantly, the four emitters are incorporated in the host CBP from high to low energy without inserting any interlayers or spacers from the exciton generation interface, that is, the interface between hole (CBP) and electron (TPBi) transport layers. This cascade design ensures that excitons are first harvested by the emitter requiring highest-energy excitons. If the order is reversed, the excitons generated would be consumed by the lower-energy emitters first, leading to insufficient amount of high-energy excitons remaining that would be able to excite the higher-energy emitters such as the blue dopant. Furthermore, no spacer regions are used here such that minimal excitons would be used by the low emission efficiency host layer for emission, and surplus energy (not delivered to the closest dopant) is allowed to transfer from high- to low-energy EMLs. The EL spectrum of this device at lighting luminance of 5000 cd/m^2 is shown in Fig. 8.5b, where the CIE coordinates are (0.37, 0.48) with a CRI of 72. The device shows a remarkable η_{EQE} and a decent efficacy at 1000 cd/m^2 of 19.2% and 28.1 lm/W, respectively. However, it is seen that even with a significantly thicker red EML, the red emission is still the weakest among the four colors, which not only results in a poor CRI, but also indicates an inferior ability of the red phosphors to harvest the generated excitons.

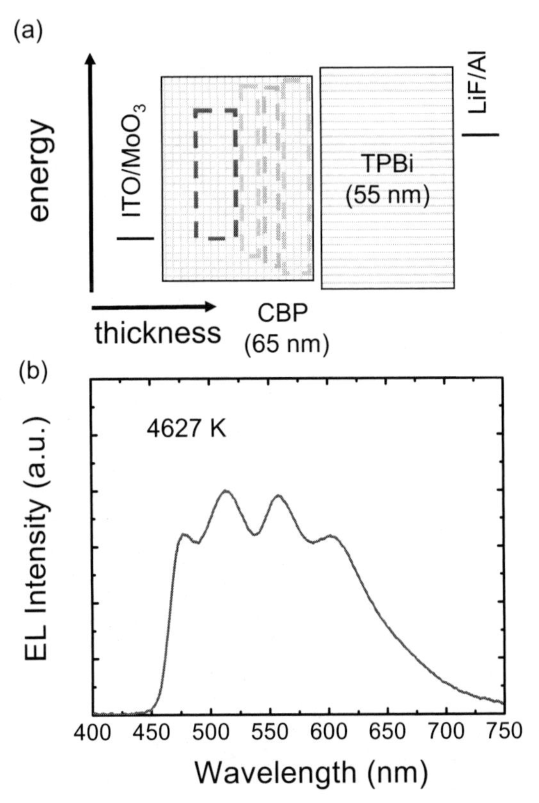

Figure 8.5 (a) Energy-level diagram and device structure of a cascade emissive layer white OLED and (b) the device EL spectrum at 5000 cd/m^2. The blue, green, yellow, and red dopants used are FIrpic, Ir(ppy)$_2$(acac), iridium (III) bis(2-phenylbenzothiozolato*N*,C$^{2'}$)(acetylacetonate)(Ir(BT)$_2$(acac)), and Ir(MDQ)$_2$(acac), respectively. This figure is generated with permission from Ref. [51]. Copyright 2012, Wiley-VCH.

8.4.3 Cascade Emissive Layers with Exciton Conversion

A further modified design to the cascaded EML concept is shown in Fig. 8.6a, where a high-energy green dopant with excellent exciton trapping capability is incorporated into the yellow and red EMLs.[51] Owing to a significant spectral overlap between the green dopant

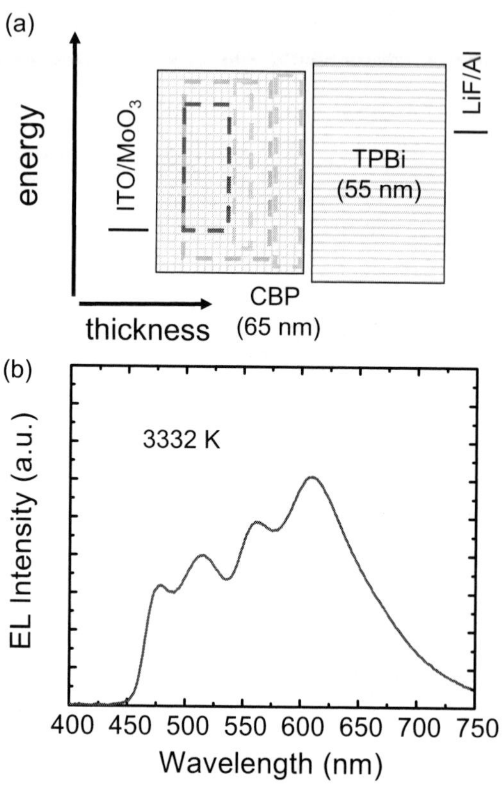

Figure 8.6 (a) Energy-level diagram and device structure of a cascade emissive layer white OLED utilizing exciton conversion and (b) the device EL spectrum at 5000 cd/m^2. The blue, green, yellow, and red dopants used are FIrpic, Ir(ppy)$_2$(acac), Ir(BT)$_2$(acac), and Ir(MDQ)$_2$(acac), respectively. This figure is generated with permission from Ref. [51]. Copyright 2012, Wiley-VCH.

emission spectrum and the red as well as yellow dopant absorption spectra, a strong exciton energy transfer from the green dopant to yellow and red dopants are expected. As a result, the spectrum of the white device shown in Fig. 8.6b reveals considerably stronger yellow and in particular red emissions, leading to a high CRI of 85 at lighting luminance of 5000 cd/m^2 with CIE coordinates of (0.44, 0.45), which corresponds to an ideal warm white illumination. This enhancement in emission intensity also yields a superior device η_{EQE} of 23.3% and efficacy of 31.0 lm/W, respectively, at 1000 cd/m^2. Such a combination of CRI of over 85 and η_{EQE} of over 20% at

5000 cd/m^2 represents the best performance in open literature to date. A hand-sized device is also demonstrated, as shown in Fig. 8.7, which illustrates the excellent color rendering capability of the light source. Further work to be made on this design include the introduction of higher triplet energy hosts such that a deeper blue dopant can be used to obtain an even higher CRI, as well as the use of higher-mobility transport layers and bipolar host materials to enhance the power efficiency. As with any all-phosphorescent white OLED designs, one of the key issues remaining is stability and lifetime the blue phosphorescent dopant, which remains an active area of research.

Figure 8.7 A white OLED on a glass panel (1 mm thick, 80 × 80 mm area) illuminating a collection of colorful candies at a luminance level of 5000 cd/m^2. Details of the device structure can be found in Ref. [51]. Excellent color rendering makes it possible to capture the diverse colors from these candies.

Chapter 9

Optical Light Out-Coupling

As discussed throughout this book, the combination of a highly reflective cathode on one end together with a semitransparent anode on the other side forms a submicron-scale optical microcavity where certain wavelengths or modes of light are enhanced and others are trapped inside depending on the total device thickness. Such a microcavity effect occurs mainly along the longitudinal, forward direction (normal to organic light-emitting diode [OLED]-emitting surface), leading to a directional- or angular-dependent emission profile. This is further exacerbated by the fact that the refractive index of organic materials and standard indium tin oxide (ITO) anode ($n \approx 1.7–1.9$) are higher than that of standard glass and plastic substrates ($n \approx 1.5$), thereby inducing even more severe total internal reflections, as shown in Fig. 9.1. These trapped modes are considered organic waveguide modes which constitute one of three major optical loss channels. Another main optical loss channel is the substrate modes which arise from refractive index mismatch between the glass/plastic substrate ($n \approx 1.5$) and the air ($n \approx 1$), resulting in considerable total internal reflections within the relatively thick substrate. The third major optical loss is attributed to surface plasmon modes taking place on the highly reflective cathode. This is induced by the close proximity between the emissive

Efficient Organic Light-Emitting Diodes (OLEDs)
Yi-Lu Chang

ISBN 978-981-4613-80-4 (Hardcover), 978-981-4613-81-1 (eBook)
www.panstanford.com

layer (EML) of the OLED and the reflective metal surface, separated only by the electron transport layer (ETL). These emissive modes are easily coupled to the surface plasmon modes, thereby forming evanescent waves that dissipate as heat. In the following, a number of the recently reported techniques to address these three types of optical out-coupling losses are presented.

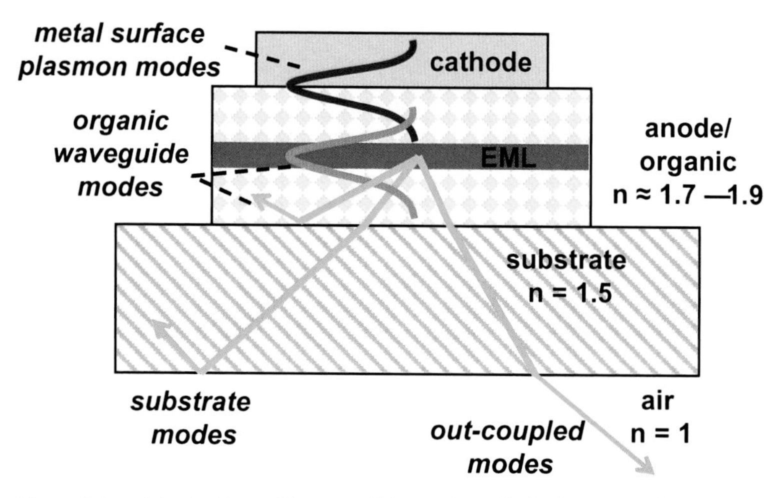

Figure 9.1 Illustration of the possible modes of light in a typical BEOLED.

A summary of the contribution of various efficiency loss channels in a standard phosphorescent OLED is shown in Fig. 9.2.[109] The electrical losses, such as energy barriers encountered during carrier injection and transport, as well as resistive losses in each organic layer could contribute to ~8% of the total loss. The nonradiative losses considering the impurity of the organic materials as well as the intrinsic emitter luminescence loss constitute to ~12% of the total loss. The absorption loss, including the reabsorption of the emitted light, is under ~5% because common organic host and transport materials have absorption bands lying well into the UV range such that the device is transparent to the emitted visible wavelengths. The last 75% of the total loss is entirely arising from optical out-coupling losses. As such, optical out-coupling is a major factor in influencing potential device efficiency improvement.

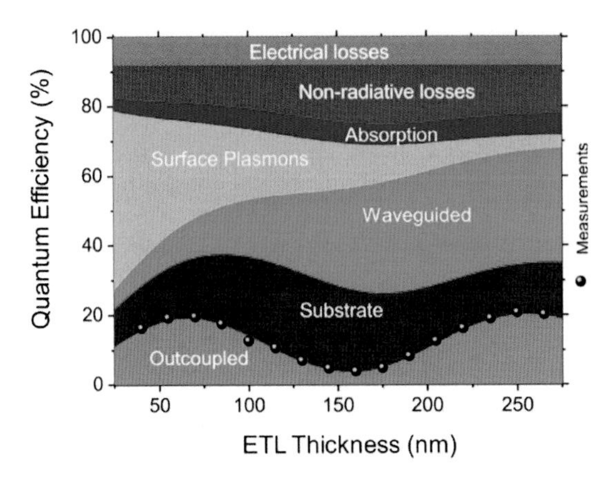

Figure 9.2 Loss channels in a standard phosphorescent BEOLED. Reproduced with permission from Ref. [109]. Copyright 2010, American Institute of Physics.

9.1 Organic Waveguide Modes

One standard approach to out-couple organic waveguide modes is to use refractive index modulation layers. Sun et al. have shown that inserting low refractive index grids between the ITO and the organic stack could extract significant waveguide modes out, as shown in Fig. 9.3.[110] Being surrounded by high-refractive-index materials, the grid material is able to redirect the light out from Snell's Law. Initially moving at a high angle to the organic light-emitting diode (OLED) normal, these modes are converted to light rays with a much smaller angle to the normal, hence entering the escape cone of the device. Compared to a standard device, this method is able to improve the external quantum efficiency by a factor of ~1.32.

Another technique was shown by Hong et al. utilizing graded index layers consisted of nanofacets, as illustrated in Fig. 9.4.[111] Since the ITO has a refractive index of ~1.9 and the glass substrate has a refractive index of ~1.5, it is possible to help guiding the light out by inserting two materials, ZrO_2 and MgO, with intermediate refractive index values of 1.84 and 1.73, respectively, in between the ITO and glass. In addition, due to the presence of the nanofacets, light exiting at high angles could still be directed back toward the normal and coupled out of the device. This approach could improve the external quantum efficiency up to ~1.19 times higher than that of a standard device. Moreover, the MgO nanofacets can be formed spontaneously without complicated fabrication processes because of the anisotropic material properties of MgO concerning its crystal orientations.

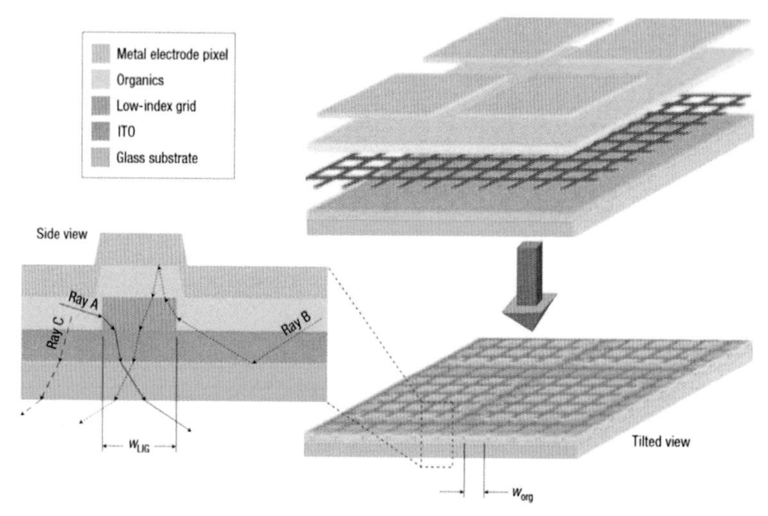

Figure 9.3 Illustration of embedded low-index grids between the substrate and the ITO anode for light extraction in a BEOLED. Reproduced with permission from Ref. [110]. Copyright 2008, Nature Publishing Group.

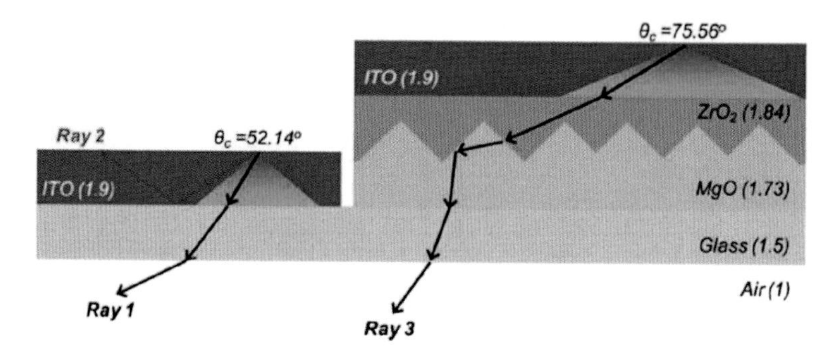

Figure 9.4 Illustration of gradient index nanofacets between the substrate and the ITO anode for light extraction in BEOLEDs. Reproduced with permission from Ref. [111]. Copyright 2010, Wiley-VCH.

Wang et al. have recently adopted a multilayer anode stack of $MoO_3/Au/Ta_2O_5$ instead of the standard MoO_3/ITO, as shown in Fig. 9.5.[112] By tuning the thickness of Au and the two high-refractive-index materials, MoO_3 and Ta_2O_5, the waveguide modes could be effectively extracted such that the external quantum efficiency reaches ~1.85 times higher than that of the standard ITO anode device. This design is also called dielectric/metal/dielectric (DMD) method, which is considered to be the closest substitute for ITO in terms of transparency and conductivity.[113] This method is also applied on a flexible plastic substrate, displaying record high $\eta_{EQE} \approx$ 40.3% for green OLEDs.

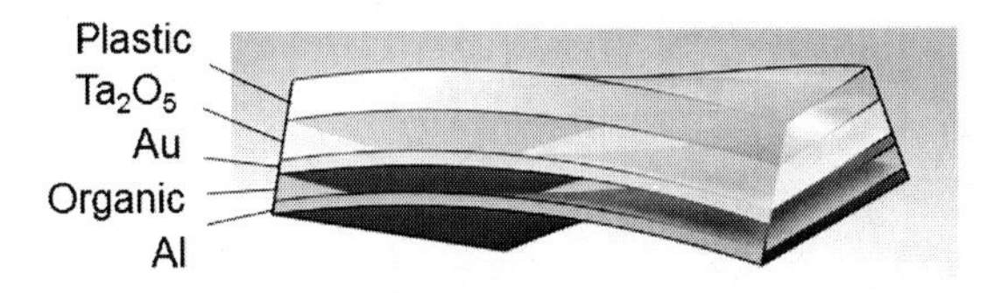

Figure 9.5 Illustration of a dielectric/metal/dielectric anode for light extraction of a BEOLED on flexible plastic. Reproduced with permission from Ref. [112]. Copyright 2011, Nature Publishing Group.

While useful, these methods do not address the effect of angular dependence of the emission profile and are insufficient for broadband emission such as a white OLED. In this regard, one approach has been demonstrated by Koo et al. using a buckling technique having a thick (~50–70 nm), corrugated structure that induces a broad periodicity distribution throughout the entire OLED stack, as shown in Fig. 9.6.[114] This subwavelength and quasiperiodic corrugated structure is able to Bragg-scatter significant organic waveguide modes out of the device. The formation of the buckling surface is due to a difference in thermal expansion coefficients of Al and polydimethylsiloxane (PDMS). After cooling down a bilayer of Al on top of thermally expanded PDMS at 100°C during Al deposition, the corrugation is formed spontaneously without further fabrication processes. This technique is shown to improve the external quantum efficiency by a factor of ~2.0 or higher across the visible range. Furthermore, the emission follows a Lambertian profile, which suggests minimal angular dependence. Very recently, Ou et al. have demonstrated a similar technique with nanopatterned substrate to produce white OLEDs with an efficiency of over 54.6%, corresponding to a efficacy of 123.4 lm/W, and a much suppressed angular dependence.[115]

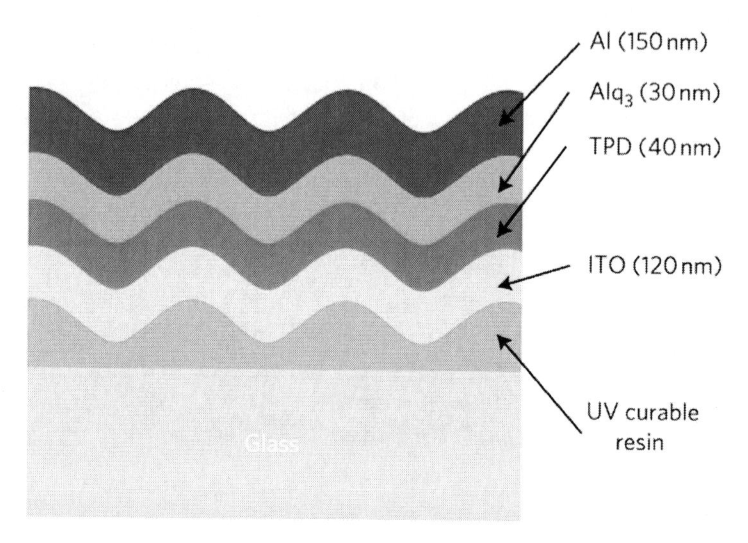

Figure 9.6 Illustration of a BEOLED with a buckling structure for light extraction. Reproduced with permission from Ref. [114]. Copyright 2010, Nature Publishing Group.

Alternatively, exotic transparent electrodes such as metal nanowires,[116] poly(3,4-ethylenedioxythiophene):poly(4-styrenesulfonate) (PEDOT:PSS),[35] and single-layer graphene[117] are probable candidates to replace ITO and minimize microcavity effects, thereby suppressing angular dependence.

9.2 Substrate Modes

Besides extracting the organic waveguide modes, a number of simple approaches have been shown to extract light trapped by the substrate modes. One simple approach is to insert a half-sphere lens with a refractive index that is well matched to that of the substrate, together with the application of an index matching gel. The lens is able to direct light trapped in the substrate out with an efficiency enhancement of up to ~1.5 across the entire visible range.[106] The main issue here is that such large optics is not ideal for practical applications either in displays or lighting. Alternatively, one can fabricate an array of microscale half-spheres (or microlenses), using standard photolithography, that are also index-matched to that of the substrate, as shown in Fig. 9.7.[110] Enhancements in external quantum efficiencies of up to ~1.8 may be attainable.[106]

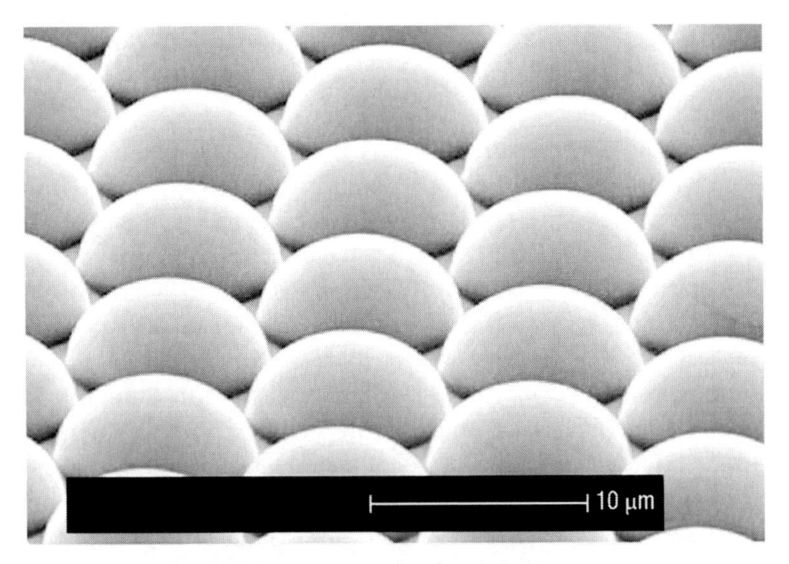

Figure 9.7 Scanning electron microscopy image of an array of microlens foil for light extraction in OLEDs. Reproduced with permission from Ref. [110]. Copyright 2008, Nature Publishing Group.

Another method is to use a transparent substrate with a higher refractive index that is better matched to that of ITO, which would minimize the refractive index difference between the anode and the substrate.[106] In this case, an enhancement in external quantum efficiency of nearly a factor of ~1.8 can be attained. However, this technique is less practical for production as high-refractive-index substrates are far more expensive than commercially available glass panels and flexible plastic substrates.

9.3 Surface Plasmon Modes

Methods to extract surface plasmon modes out of an OLED device are fairly limited. As shown in Fig. 9.2, less surface plasmon modes are present when the thickness of the ETL is increased so that the EML is farther away from the highly reflective cathode. However, as the length of the ETL increases, the total device thickness also increases, which in turn results in an increase in the organic waveguide modes, as shown in Fig. 9.2.[109] Additionally, increasing the ETL thickness would require chemical doping using electron-donating organic molecules to improve the conductivity, and compensate for the increased distance that an electron would have to travel to reach the host.[67]

Another way to suppress the surface plasmon modes is to employ emitters that are predominantly horizontally oriented (i.e., parallel to the OLED emitting surface).[118] In general, the emission of an arbitrarily oriented dipole contains three orthogonal orientations, ||TE, ||TM, and ⊥TM, where TE represents transverse electric mode and TM denotes transverse magnetic mode, respectively. Because the emitter dipoles normal (perpendicular) to the emitting surface have a higher likelihood to couple to the metal surface plasmon modes, an emitter with primarily horizontal (parallel) dipole orientation would have a lower number of modes available to be coupled to the surface plasmon modes compared to that of an emitter having an isotropic orientation (67% parallel and 33% perpendicularly oriented). This means more modes can be extracted out of the device from a predominantly horizontally oriented dipole emitter. Indeed, it has been shown that an OLED with a green phosphorescent emitter having 76% parallel dipole orientation could attain an external quantum efficiency of ~30%.[11] From optical simulations, the same emitter with 100% parallel orientation could give a possible external quantum efficiency of up to ~45%. The obvious caveat with this approach is the difficulty in controlling the emitter orientation during device fabrication.

Chapter 10

Stability and Degradation

After discussing highly efficient organic light-emitting diode (OLED) architectures in detail, it is appropriate to describe device operational stability and material degradation processes after prolonged device operation. In organic semiconductors, although there are no dangling bonds that introduce defects as in inorganic semiconductors, the molecule stability and the interface between two different layers of organic material are critical due to differences in surface energy, dipole moment, exciton formation/migration rate, and carrier transport property. These aspects are crucial to device performance over time and central to industrial applications of OLED technology.

10.1 Efficiency Roll-Off

Organic light-emitting diode (OLED) devices show a characteristic efficiency roll-off (efficiency droop) under high current densities (high luminance levels), as shown in Fig. 2.1b. This is detrimental for high luminance applications such as general illumination sources that require ~3000–5000 cd/m^2 luminance. In standard phosphorescent devices, the external quantum efficiency typically drops to ~50%–75% of its peak value at a luminescence of ~500

Efficient Organic Light-Emitting Diodes (OLEDs)
Yi-Lu Chang

ISBN 978-981-4613-80-4 (Hardcover), 978-981-4613-81-1 (eBook)
www.panstanford.com

cd/m^2, a brightness suitable for a flat-panel display. There are a number of physical processes responsible for such efficiency roll-off. The major processes include triplet–triplet annihilation (TTA), triplet–polaron annihilation (TPA), and drift in the charge carrier balance (CCB).

10.1.1 Triplet–Triplet Annihilation

TTA is one major quenching process in phosphorescent OLED devices.[16] This process is traditionally based on the Dexter energy transfer mechanism described in Process [2.7]. When the triplet energy of the host is considerably higher than that of the guest without significant back energy transfer from the guest to the host, the predominant TTA process may occur between two guest molecules. However, if phosphorescent guest molecules are used, a strong spin–orbit coupling of the guest molecules may further enhance the TTA process via a long-range Förster mechanism.[17] In the case when relatively lower-triplet-energy host materials are used, the triplet excitons' (triplets) TTA among the host molecules could be significant. This would reduce the amount of host triplets that may be energy-transferred to the guests to provide more photons. In addition, since exciton density rises with increasing electrical current injection, it can be expected that under high current injection, the triplets on the guest are saturated and the amount of host triplets becomes sufficiently high so that TTA could even take place between host and guest molecules. One method to reduce TTA is to employ high-triplet-energy host materials and select guest phosphors with a short triplet lifetime so that most guest triplets would be emitted radiatively in a short period of time before significant accumulation in the system.

10.1.2 Triplet–Polaron Annihilation

TPA is another major quenching process in OLEDs. This process can be expressed as

$$T_1 + e^- \rightarrow S_0 + e^{-*} \qquad [10.8]$$

$$T_1 + h^+ \rightarrow S_0 + h^{+*}, \qquad [10.9]$$

where charged species (or polarons) are represented by electrons (e^-) and holes (h^+), respectively. These processes describe the annihilation of triplet excitons with free or trapped charge carriers by a Förster-type mechanism[119] to promote the charged species into excited states (e^{-*} and h^{+*}). As mentioned above, the density of triplet excitons in the device becomes extremely high under high current densities particularly near the exciton formation regions of a phosphorescent device. For the charged species, the density and location of accumulated charges depend on organic–organic or organic–metal heterojunction interface where the energy barrier offsets are high.[120] The proportion of the accumulated electrons and holes, in turn, depends on the electron/hole mobilities of the host and transport layers. One way to minimize TPA is then to construct devices where the exciton formation region is spatially apart from the carrier accumulating interfaces or simply eliminate the number of organic transport layers in the device.

10.1.3 Charge Carrier Balance

The CCB is a critical parameter influencing the device efficiency, as shown in Eq. 2.2. Any drift in the CCB during device operation would result in a waste of charge carriers and thus contribute to the efficiency droop. The CCB is considered a dynamic process depending on densities of each carrier type under different injection levels.[17] This is due to multiple factors such as the difference in mobility and in injection barrier heights encountered by electrons and holes. The best approach to balance charge carriers is to optimize the electron and hole transport layer thickness as well as employ ambipolar host materials that offer comparable electron and hole mobilities.

10.2 Material Degradation

While efficiency droop concerns with the efficiency value at practical brightness levels, material degradation determines the overall device lifetime. Major sources of degradation include cathode oxidation, anode degradation, electromigration, molecular aggregation, and molecular fragmentation.

10.2.1 Cathode Oxidation

Low-work-function cathode materials are extremely reactive to oxygen and moisture. These reactions lead to the formation of insulating metal oxides, which appear as nonemissive areas known as dark spots.[121] These dark spots will grow with time as the reaction continues. In fact, it has been shown that by peeling off a delaminated Al cathode, and redepositing a fresh layer of Al, the device still performs well.[121] Fortunately, using state-of-the-art encapsulation methods like a metal or glass cap coverage, such corrosive cathode oxidation can be eliminated for majority of the products. In terms of flexible OLEDs, however, effective device encapsulation remains a challenging engineering problem.

10.2.2 Anode Degradation

For commercial application, indium tin oxide (ITO) anodes have been used for the last two decades in standard bottom-emission organic light-emitting diodes (OLEDs). To increase the work function of ITO, UV-ozone, or oxygen plasma treatment is typically applied to the ITO surface before OLED deposition. Such surface treatment forces oxygen atoms closer to the surface of the ITO. These oxygen atoms are unstable and are prone to diffusion into the adjacent organic layers thereby reducing the work function of the anode.[121] It has also been realized that the UV-ozone-treated ITO surface serves as an oxygen reservoir that has a tendency to interact and possibly shift the highest occupied molecular orbital (HOMO) level of the adjacent organic layer away from the Fermi level, thereby producing a gap state in the energy gap of the organic that becomes a charge recombination center.[122] One method to avoid this issue is to insert one or more extra hole injection layers (e.g., MoO_3 or HATCN) to stabilize the device.

10.2.3 Electromigration

Another degradation factor involves electric field–induced migration of metal atoms and organic species. It has been known for a long time from traditional inorganic semiconductors that metal cations are prone to diffusion in a device under an electric field across a *p-n* junction. Similarly in an OLED, metal cations from ITO anode have been found to diffuse into the organic stack during operation.[121] Such diffusion of metal ions into the bulk organic stack forms fixed deep-level charge traps or exciton-quenching centers.[120] For charge transport organic layers, negatively or positively charged molecules may also migrate under an applied electric field. As a result, the composition and electrical properties of the organic stack will deviate from the initial design, leading to nonoptimal properties.

10.2.4 Molecular Aggregation

Organic molecules (typically hosts and transport layers) are prone to reorganization in the solid state to form compact, crystalline structures when injected with energy.[123] This energy can be in the form of heat, high-energy (ultraviolet) photoexcitation, and electrical excitation that involves high-energy excitons and polarons.[123] Such molecule aggregation is signified by the presence of a shoulder peak that is red-shifted from its primary emission.[123] The shoulder peak directly indicates the formation of quenching states that inevitably reduces the device efficiency since energy back transfer from the dopants to the host or transport layer becomes more probable. Incidentally, it was found that the wider the energy gap (higher energy) a molecule has, the more severe its aggregation becomes.[123] In some cases, by merely thermal evaporation without driving the device, aggregation could occur due to thermal activation during deposition. At the same time, as mentioned in previous chapters, high-energy hosts and transport layers are critical for confining the exciton delivery to the dopants. There is therefore a trade-off between the selection of less aggregation-prone materials and materials that exhibit a high-energy gap.

10.2.5 Molecular Fragmentation

Organic molecular fragmentation may occur on the guest, host, and charge transport layers and, in particular, molecules near organic–organic and organic–metal interfaces. This could be triggered by high-energy exciton and polaron accumulation, repeated high-energy electrical excitation and aging, or UV light illumination–induced photoaging.[20] Under these processes, high-energy excited states of the organic molecule are accessed which exceed the deactivation energy of the relatively weaker chemical bonds such as C–N, C–metal, N–metal, or C–P bonds.[20] Upon deactivation, these weaker bonds are ruptured at an excited state, which results in a series of radical addition reactions to form stabilized radicals. The accumulation of the neutral radical species and their reduced and/or oxidized forms in the device would certainly become nonradiative recombination centers and luminescence quenchers. One approach to address this issue is to design more rigid complexes that have minimal freedom for the rearrangement of excited molecule, thereby preventing dissociation of the weaker bonds.[20] For the guest emitters, because the reason for molecule degradation often involves excitons, a reduction in device exciton density by using triplet emitters with a short triplet lifetime may alleviate the degradation process. Alternatively, as mentioned previously, thermally activated delayed fluorescence (TADF) and exciplex-forming hosts may be used to reduce the amount of triplets during device operation, while still utilizing them usefully by up-converting to singlets. In addition, the interface between organics and metal (cathode and anode) is another probable place for molecular breakage, which would lead to poor carrier injection, hence lowering device efficiency over time. One probable way to resolve this issue is to employ a narrow-energy-gap, more stable transport layer on both cathode and anode sides such that the excitons diffused to (and accumulated at) either electrode will possess considerably smaller energy.

Chapter 11

Applications in Displays and Lighting

11.1 Displays

The organic light-emitting diode (OLED) has been considered by many as the ultimate display technology. Indeed it is already dominating the portable displays industry and is quickly growing in the flat-panel displays arena. Due to its uniquely flexible, thin form factor, exotic types of displays such as wearable displays have also been introduced by major electronic companies like Apple and Samsung.

For small displays, to achieve high resolution with high pixel density, the prevailing approach has been the use of top-emission organic light-emitting diodes (TEOLEDs) (see Chapter 7) such as the one introduced by Sony, named super top-emission OLEDs shown in Fig. 11.1. These OLEDs are fabricated directly on top of the active matrix transistors such that minimal nonemissive area (black matrix) is taken by the transistors.

It can be seen that each pixel is consisted of three OLED subpixels of the primary colors. Notice that the thickness of each subpixel is being optimized individually for each color by taking advantage of the microcavity effect to maximize light out-coupling. This would enhance the emission intensity of each color so that less power is consumed to achieve a desired brightness and higher display contrast can be achieved. In addition, color filters are added to enhance the

Efficient Organic Light-Emitting Diodes (OLEDs)
Yi-Lu Chang

ISBN 978-981-4613-80-4 (Hardcover), 978-981-4613-81-1 (eBook)
www.panstanford.com

color purity of each primary color. Currently, most portable displays from major players like Samsung Display are employing TEOLED technology similar to Fig. 11.1.

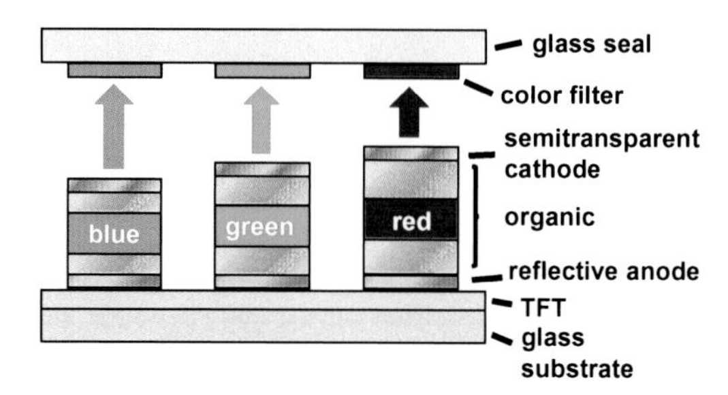

Figure 11.1 Schematic diagram of a top-emission AMOLED structure employed by Sony Corporation.

For larger-sized displays with a longer viewing distance, to achieve the high resolution required, the pixel density needed is actually not nearly as high as in the small portable displays case. Hence, standard bottom-emission organic light-emitting diodes (BEOLEDs) will suffice, as shown in Fig. 11.2.[124]

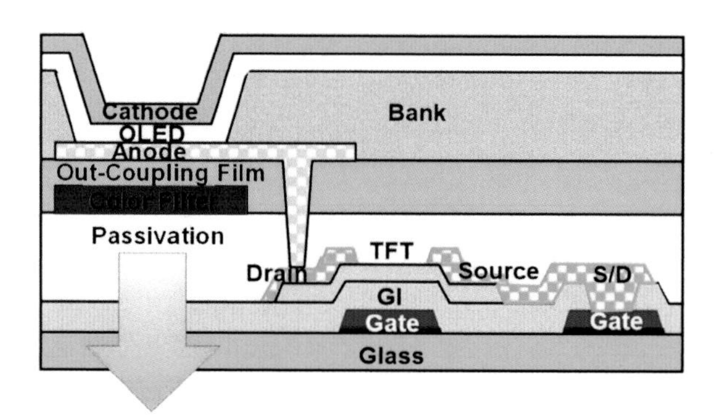

Figure 11.2 Schematic diagram of an AMOLED design adopted by LG Display. Reproduced with permission from Ref. [124]. Copyright 2013, Wiley-VCH.

In this respect, there are currently two competing color patterning designs thus far, as shown in Fig. 11.3. The first one is called red, green, blue (RGB), side-by-side pixel design where each of the three primary color subpixels are placed adjacent to one another with the blue subpixel being 1.5 times as large as the green and red subpixels. The reason for a larger blue pixel is that state-of-the-art blue emitter used is a fluorescent type that is not as efficient as the phosphorescent green and red emitters. Note, once again the length of each primary color OLEDs are optimized for maximum forward light output for each color. Here, no color filters are needed, and polarizers are used to minimize ambient reflections. This design is currently being adopted by Samsung Display, as shown in Fig. 11.3a. This approach has the advantage of high efficiency and lower power consumption. However, one major issue is the yield of the fabrication process when the size of the display becomes large. This is primarily because the fine metal mask (FMM) used for pixel patterning becomes heavy after scaling up and are prone to sagging, which leads to shadowing effect and misalignment of the pixels. Further improvement in the fabrication process was made by Samsung Display to utilize a small-mask scanning (SMS) technique for the OLED deposition. This method is a similar to FMM; however, instead of the substrate and the mask being fixed in one position as with the FMM process, the SMS method, the mask is held constant and the substrate moves. As a result, the mask is only as large as the panel and the possibility of sagging is minimized. However, the yield is still not sufficiently high; hence the production cost remains significant. Other alternative OLED pixel patterning techniques are shown in Table 11.1.

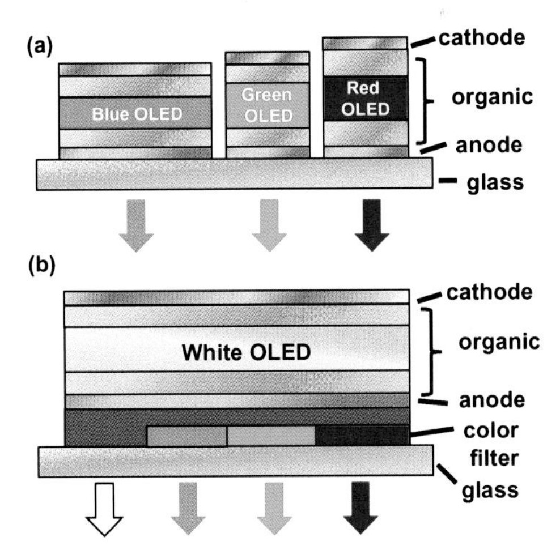

Figure 11.3 Illustration of a pixel in an AMOLED using (a) RGB side-by-side color patterning that is adopted by Samsung Display (b) and using RGBW color patterning that is adopted by LG Display.

Alternatively, it is possible to employ another color patterning technique using red, green, blue, and white subpixels (RGBW) with color filters. This technique, originally developed by Kodak, has been adopted by LG Display, as shown in Fig. 11.3b. Each pixel has an identical white OLED, and color filters are used to create red, green and blue pixels. Note that each color filter blocks nearly two-thirds of the emission from the white OLED, thereby reducing the overall efficiency. To compensate for this, an additional white pixel without a color filter is used to boost the brightness and efficiency of the overall pixel group. There is only a single white OLED structure in each pixel that consists of a tandem structure of a blue unit plus a mixture of green and red unit, which lowers the driving current, and prolongs device lifetime, as shown in Fig. 11.4. This has the advantage of easier fabrication (higher volume and yield) as the high density color filters patterning are already well developed from decades of LCD display fabrication. In addition, with the advances of metal–oxide thin-film transistor (TFT) backplanes having higher mobility and improved uniformity across a larger area via circuit compensation, it is possible to reduce the black matrix occupied by the transistors, increase the density of emissive pixels and accommodate for larger display sizes. Indeed, LG Display was the first to introduce 65" 4K resolution OLED TVs for mass production, which are arguably the highest-quality flat-panel display to date.

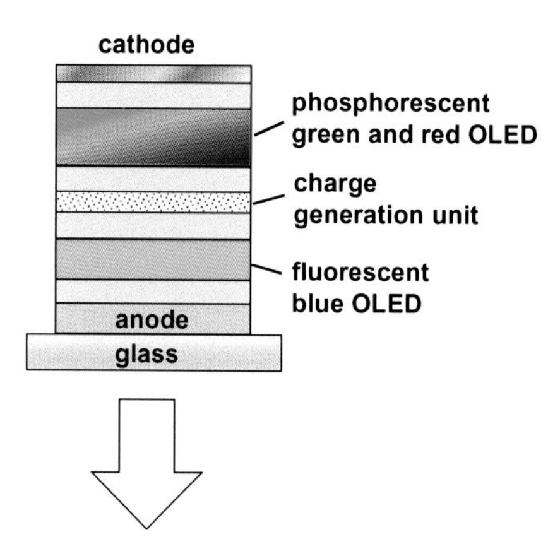

Figure 11.4 Tandem white OLED employing a blue fluorescent OLED in one unit and a mixed green plus red phosphorescent OLED in a second unit. This design is adopted by LG Display as the white OLED in each pixel of the OLED TV panel.

Regardless of the color-patterning techniques, there remains an issue with OLED differential aging. In the RGBW design, each white OLED is essentially made up of different primary color emissive layers (EMLs) (an EML consisted of green and red dopants, together with a separate blue EML), where each primary color emitter could be subjected to a different electrical aging time. As a result, the white OLED emission may become distorted, giving a nonideal color coordinate for each pixel compared to the initial setting. Similarly, in the RGB side-by-side design, each primary color subpixel will experience different aging time depending on a particular usage style. This will result in certain subpixels become darker than others, leading to a burn-in effect on the display. One way to compensate this is to adopt a more sophisticated driving circuitry to correct the color shifts (e.g., increase current density for the severely aged subpixel to match its luminance to those of other less aged subpixels).

11.2 Solid-State Lighting

In terms of solid-state lighting, a number of companies worldwide (LG Chemicals, Philips, OSRAM, Merck, etc.) are actively trying to enter the market by driving down the cost and producing efficient, long-lasting OLEDs in order to compete with inorganic LEDs, compact fluorescent lamps, and incandescent bulbs. Compared to incandescent bulbs, which are inefficient point sources that need to be shielded to protect against both glare and heat, OLEDs are surface sources that exhibit very little glare and heat. OLEDs are also superior to fluorescent tubes which are prone to yellowing, noise production, and environmental damage with the use of toxic mercury. In contrast to inorganic LEDs, OLEDs do not need additional diffusers, light guides, heat sinks, and other components typical of LEDs.

For general lighting, the desired emission is of a soft Lambertian pattern with minimal angular dependence such that a BEOLED is sufficient. In this regard, state-of-the-art lighting panels from major players such as Philips are typically made of tandem bottom-emitting structures with each repeating units similar to LG's white OLED pixel in their RGBW displays shown in Fig. 11.4. As discussed earlier, such employment of multiple repeating white units has the advantage of current recycling, or generating multiple photons

Table 11.1 Comparison of various methods of fabricating OLED displays

	FMM	LITI	SMS	Printing	RGBW
Materials	Invar, SUS	YAG laser, Donor film	Invar, SUS	No Mask	No Mask
Difficulty	Low	High	Low	High	Low
Cost	High	High	High	Low	Low
PPI	~250 ppi	>300pi	~250 ppi	~250 ppi	>300 ppi
Performance	High	High	High	Low	High
Deposition	Heat in vacuum	Heat in vacuum	Heat in vacuum	No Heat in N_2	Heat in vacuum
Pros	Already in production, long organic material lifetime	Supports high ppi	Supports large size	Good material utilization	High yields, high ppi, long organic material lifetime
Cons	Substrate sagging from heat, poor material utilization	Low evaporation rate, poor material utilization	Poor material utilization, ppi limited	Slow new material introduction, ppi limited by nozzle size	Reduced light output, poor material utilization
Substrate size	Gen 5.5	Gen 5.5	Gen 8	Gen 8	Gen 8

FFM, LITI, SMS stand for finite metal mask, laser induced thermal imaging, and small mask scanning, respectively.

Source: Data is acquired from the OLED Association.

Table 11.2 Reported laboratory white OLED panel performance in 2013–2014

Company	Luminance (cd/m^2)	Efficacy (lm/W)	Area (cm^2)	CRI (Ra)	CCT (K)	L_{70} (1000 hours)	Drive (V)	Structure
Konica	1,000	131	15	82	2800	27.5		
Minolta	3,000	118						
SEL/Sharp	1,000	113	81		3270	400	8	Triplet Stack
	5,000	105					8.4	
Panasonic	1,000	110	25	81	2600	40	5.5	Double Stack
	3,000	98				10	6.0	(Tandem)
UDC	1,000	780	20	85	3030	165	7.1	Triplet Stack
	3,000	60		86	2880	25	7.8	
LG Chem	3,000	82	76	84	2900	30	8.5	Triplet Stack
CDT/	1,000	56	13	80	2900		4.3	Single Stack, solution
Sumitomo	3,000	48		82			4.8	processed up to EML

Source: Data is obtained from the OLED Association.

(depending on the number of units) for each electron–hole pair injected. This significantly lowers the current density needed to achieve a desired brightness, at the cost of requiring higher driving voltages (each unit is connected in series), while keeping the overall power consumption low enough using electrical doping technology. Since OLED degradation is critically dependent on current density, this approach aims at prolonging the lifetime of OLEDs. The major caveat of this method is the tedious, multilayer fabrication processes that involve precise doping, which drives down the reproducibility and yield, thereby preventing the cost of each panel from dropping considerably.

Table 11.2 shows a list of prototype OLED panels that have been demonstrated recently. It is clear that most panels are based on stacked OLED design with the number of repeating units indicated by the driving voltage required. Although these high-efficiency prototypes are quite promising, other critical parameters such as lifetime, cost, and color quality, as well as form factor also need to be considered collectively. Currently, the extraction or out-coupling efficiency has the most potential for improvement. Other than this, the next major challenge does not concern with the efficiency, but it has to do with fabrication yield and cost, as well as overall panel lifetime.

Chapter 12

Conclusions and Outlook

To summarize, in an organic light-emitting diode (OLED), the internal quantum efficiency (η_{IQE}) is dependent on two main factors. The first factor is to obtain a balanced flow of electrons and holes into the emission layer. The second factor concerns with the fraction of recombining electron–hole pairs that results in the production of visible photons. Typically, it is convenient to utilize a host that controls charge transport and exciton energy transfer, together with a dopant that receives the excitonic energy and releases it in the form of photons. Exciton-harvesting dopants such as phosphorescent molecules have been the dopant of choice for green and red emissions with adequate efficiency and lifetime. However, the lifetime of blue phosphorescent emitters remains unsatisfactory. Hence, most panel manufacturers utilize a hybrid system of phosphorescent green and red dopants together with a stable blue fluorescent dopant, which yields a system η_{IQE} of around 75%.

One inherent issue with the use of phosphorescent molecules is that their excited-state energy stays for a much longer time (microseconds) than in fluorescent molecules (nanoseconds). Such high excitonic energy accumulating in the device can trigger undesirable quenching processes that not only lower the η_{IQE}, but also damage the device, especially the delicate organic–organic and organic–metal heterojunction interfaces. As a result, phosphorescent OLEDs often display more rapid degradation

Efficient Organic Light-Emitting Diodes (OLEDs)
Yi-Lu Chang

ISBN 978-981-4613-80-4 (Hardcover), 978-981-4613-81-1 (eBook)
www.panstanford.com

under high luminance operation. Recently, fluorescent emitters having small singlet–triplet energy-level differences, are found to promote efficient thermally activated up-conversion of triplet to singlet states, leading to delayed fluorescence and a high η_{IQE} of ~100%. However, further development is needed to verify such low-cost, high-efficiency system can exhibit a sufficiently long lifetime.

In terms of device architecture, a number of advanced methods have been introduced to improve the η_{IQE} of fluorescent and phosphorescent blue dopants, including the use of exciton harvesting and conversion dopants, as well as the employment of thermally activated delayed fluorescence (TADF) hosts and exciplex-forming cohosts. At the heart of these concepts is the minimization of nonemissive host triplet states via reverse intersystem crossing (RISC) and energy-level matching between host donors and dopant acceptors. Regardless of the design, the most fundamental requirement is triplet exciton confinement to the emissive layer (EML), and facile host energy transfer to dopants, which are preferably capable of harvesting both triplets and singlets to contribute to radiative emission. Currently, although the efficiency of these concepts show great promise, the lifetime remains to be proven with the goal of out-performing state-of-the-art blue fluorescent and phosphorescent OLEDs.

The light out-coupling efficiency in an OLED is defined by the ratio of visible photons emitted from the panel to the photons actually generated in the EML. The major loss in light out-coupling is the trapping of photons in the two electrodes, the transparent substrate, and the organic layers. Due to the refractive index mismatch between the organic materials, the substrate, and air, there exists a small cone of incidence where light can be out-coupled. Despite this challenge, a number of light out-coupling enhancement techniques can be used to improve the light extraction efficiency. These include exploiting microcavity effect so that the desired wavelength of light is emitted predominantly in the normal (forward) directions of the panel, bending the light toward the normal using microlens arrays or periodic/aperiodic nanostructures, employing exotic transparent conducting electrodes, and incorporating scattering centers or rough interfaces such that light makes numerous attempts to escape at different angles. While most of these techniques are effective at improving the performance in the laboratory setting, only a few methods shown are compatible with inexpensive manufacturing at a large scale. It is worth emphasizing that once a cost-effective solution for light out-coupling is attained, the lifetime of the product will also be naturally extended due to the requirement of less driving current to achieve a desired brightness.

In the short term, most competitive OLED products will be made primarily by vacuum deposition techniques since solution processing still yields devices with unsatisfactory efficiency and lifetime. As shown in Table 11.2, even with a hybrid of vacuum deposition and solution processing, the performance is still only half of those fabricated entirely by vacuum deposition. The inability to make reliable and efficient OLEDs by solution processing will certainly prevent a rapid cost reduction. Roll-to-roll processing on flexible substrates is another attractive, low-cost approach that is currently being pursued by Konica Minolta. The challenge here lies in the development of a reliable barrier technology that prevents water and oxygen penetration through the flexible plastic substrates and covers. Eventually, vacuum processing on flexible substrates may yield many more intriguing products, which are on the short-term roadmap of major electronic display companies such as Samsung Display and LG Display.

After observing a remarkable technological progress over the past two decades, OLED technology is steadily becoming the ultimate display technology of the coming decades. As an efficient, design-friendly broadband light source, OLED will likely provide a niche application initially before eventually excelling in the exploding solid-state lighting market.

References

1. Tang CW, VanSlyke SA (1987) Organic electroluminescent diodes. *Appl. Phys. Lett.* **51**, 913–915.
2. Baldo MA, O'Brien DF, You Y, Shoustikov A, Sibley S, Thompson ME, Forrest SR (1998) Highly efficient phosphorescent emission from organic electroluminescent devices. *Nature* **395**, 151–154.
3. Uoyama H, Goushi K, Shizu K, Nomura H, Adachi C (2012) Highly efficient organic light-emitting diodes from delayed fluorescence. *Nature* **492**, 234–238.
4. Kido J, Kimura M, Nagai K (1995) Multilayer white light-emitting organic electroluminescent device. *Science* **267**, 1332–1334.
5. Minaev B, Baryshnikov G, Agren H (2014) Principles of phosphorescent organic light emitting devices. *Phys. Chem. Chem. Phys.* **16**, 1719–1758.
6. Scholes GD (2003) Long-range resonance energy transfer in molecular systems. *Annu. Rev. Phys. Chem.* **54**, 57–87.
7. Dexter DL (1953) A theory of sensitized luminescence in solids. *J. Chem. Phys.* **21**, 836–850.
8. Okada S, Okinaka K, Iwawaki H, Furugori M, Hashimoto M, Mukaide T, Kamatani J, Igawa S, Tsuboyama A, Takiguchi T, Ueno K (2005) Substituent effects of iridium complexes for highly efficient red OLEDs. *Dalton Trans.* 1583–1590.
9. Wang ZB, Helander MG, Qiu J, Puzzo DP, Greiner MT, Liu ZW, Lu ZH (2011) Highly simplified phosphorescent organic light emitting diode with >20% external quantum efficiency at >10,000 cd/m^2. *Appl. Phys. Lett.* **98**, 073310.
10. Adachi C, Baldo MA, Thompson ME, Forrest SR (2001) Nearly 100% internal phosphorescence efficiency in an organic light-emitting device. *J. Appl. Phys.* **90**, 5048–5051.

11. Kim K-H, Moon C-K, Lee J-H, Kim S-Y, Kim J-J (2014) Highly efficient organic light-emitting diodes with phosphorescent emitters having high quantum yield and horizontal orientation of transition dipole moments. *Adv. Mater.* **26**, 3844–3847.

12. Chang YL, Lu ZH (2013) White organic light-emitting diodes for solid-state lighting. *J. Disp. Technol.* **9**, 459–468.

13. Hofmann S, Thomschke M, Lüssem B, Leo K (2011) Top-emitting organic light-emitting diodes. *Opt. Express* **19**, A1250–A1264.

14. Reineke S, Thomschke M, Lüssem B, Leo K (2013) White organic light-emitting diodes: status and perspective. *Rev. Mod. Phys.* **85**, 1245–1293.

15. Singh J (2007) Radiative recombination and lifetime of a triplet excitation mediated by spin-orbit coupling in amorphous semiconductors. *Phys. Rev. B* **76**, 085205.

16. Baldo MA, Adachi C, Forrest SR (2000) Transient analysis of organic electrophosphorescence. II. Transient analysis of triplet-triplet annihilation. *Phys. Rev. B* **62**, 10967–10977.

17. Murawski C, Leo K, Gather MC (2013) Efficiency roll-off in organic light-emitting diodes. *Adv. Mater.* **25**, 6801–6827.

18. Lamansky S, Djurovich P, Murphy D, Abdel-Razzaq F, Lee HE, Adachi C, Burrows PE, Forrest SR, Thompson ME (2001) Highly phosphorescent bis-cyclometalated iridium complexes: synthesis, photophysical characterization, and use in organic light emitting diodes. *J. Am. Chem. Soc.* **123**, 4304–4312.

19. Xiao LX, Chen ZJ, Qu B, Luo JX, Kong S, Gong QH, Kido JJ (2011) Recent progresses on materials for electrophosphorescent organic light-emitting devices. *Adv. Mater.* **23**, 926–952.

20. Schmidbauer S, Hohenleutner A, Konig B (2013) Chemical degradation in organic light-emitting devices: mechanisms and implications for the design of new materials. *Adv. Mater.* **25**, 2114–2129.

21. Jankus V, Chiang C-J, Dias F, Monkman AP (2013) Deep blue exciplex organic light-emitting diodes with enhanced efficiency; p-type or e-type triplet conversion to singlet excitons? *Adv. Mater.* **25**, 1455–1459.

22. Kondakov DY, Pawlik TD, Hatwar TK, Spindler JP (2009) Triplet annihilation exceeding spin statistical limit in highly efficient fluorescent organic light-emitting diodes. *J. Appl. Phys.* **106**, 124510.

23. Zhang Q, Li B, Huang S, Nomura H, Tanaka H, Adachi C (2014) Efficient blue organic light-emitting diodes employing thermally activated delayed fluorescence. *Nat. Photon.* **8**, 326–332.

24. Zhang D, Duan L, Li C, Li Y, Li H, Zhang D, Qiu Y (2014) High-efficiency fluorescent organic light-emitting devices using sensitizing hosts with a small singlet–triplet exchange energy. *Adv. Mater.* **26**, 5050–5055.

25. Kohler A, Bassler H (2009) Triplet states in organic semiconductors. *Mater. Sci. Eng. R* **66**, 71–109.

26. Förster T (1948) Intermolecular energy migration and fluorescence. *Ann. Phys.* **2**, 55–75.

27. Braslavsky SE, Fron E, Rodriguez HB, Roman ES, Scholes GD, Schweitzer G, Valeur B, Wirz J (2008) Pitfalls and limitations in the practical use of forster's theory of resonance energy transfer. *Photochem. Photobiol. Sci.* **7**, 1444–1448.

28. Chang Y-L, Kamino BA, Wang Z, Helander MG, Rao Y, Chai L, Wang S, Bender TP, Lu Z-H (2013) Highly efficient greenish-yellow phosphorescent organic light-emitting diodes based on interzone exciton transfer. *Adv. Funct. Mater.* **23**, 3204–3211.

29. Pope M, Swenberg CE (Eds.), *Electronic Processes in Organic Crystals* (Oxford University Press, New York, 1999).

30. Kohler A, Bassler H (2011) What controls triplet exciton transfer in organic semiconductors? *J. Mater. Chem.* **21**, 4003–4011.

31. Lunt RR, Giebink NC, Belak AA, Benziger JB, Forrest SR (2009) Exciton diffusion lengths of organic semiconductor thin films measured by spectrally resolved photoluminescence quenching. *J. Appl. Phys.* **105**, 053711.

32. Fick A (1995) On liquid diffusion (reprinted from *London, Edinburgh, and Dublin Philosophical Magazine and Journal of Science*, Vol. 10, p. 30, 1855). *J Membr. Sci.* **100**, 33–38.

33. Wunsche J, Reineke S, Lussem B, Leo K (2010) Measurement of triplet exciton diffusion in organic light-emitting diodes. *Phys. Rev. B* **81**, 245201.

34. Greiner MT, Helander MG, Tang WM, Wang ZB, Qiu J, Lu ZH (2012) Universal energy-level alignment of molecules on metal oxides. *Nat. Mater.* **11**, 76–81.

35. Cai M, Ye Z, Xiao T, Liu R, Chen Y, Mayer RW, Biswas R, Ho K-M, Shinar R, Shinar J (2012) Extremely efficient indium–tin-oxide-free green

phosphorescent organic light-emitting diodes. *Adv. Mater.* **24**, 4337–4342.

36. Liao LS, Klubek KP (2008) Power efficiency improvement in a tandem organic light-emitting diode. *Appl. Phys. Lett.* **92**, 223311.
37. Helander MG, Wang ZB, Qiu J, Greiner MT, Puzzo DP, Liu ZW, Lu ZH (2011) Chlorinated indium tin oxide electrodes with high work function for organic device compatibility. *Science* **332**, 944–947.
38. Murgatro Pn (1970) Theory of space-charge-limited current enhanced by Frenkel effect. *J. Phys. D: Appl. Phys.* **3**, 151.
39. Scott JC, Malliaras GG (1999) Charge injection and recombination at the metal-organic interface. *Chem. Phys. Lett.* **299**, 115–119.
40. Helander MG, Greiner MT, Wang ZB, Lu ZH (2010) Pitfalls in measuring work function using photoelectron spectroscopy. *Appl. Surf. Sci.* **256**, 2602–2605.
41. Helander MG, Wang ZB, Greiner MT, Qiu J, Lu ZH (2009) Experimental design for the determination of the injection barrier height at metal/organic interfaces using temperature dependent current-voltage measurements. *Rev. Sci. Instrum.* **80**, 033901.
42. Su S-J, Gonmori E, Sasabe H, Kido J (2008) Highly efficient organic blue- and white-light-emitting devices having a carrier- and exciton-confining structure for reduced efficiency roll-off. *Adv. Mater.* **20**, 4189–4194.
43. Sasabe H, Takamatsu J-I, Motoyama T, Watanabe S, Wagenblast G, Langer N, Molt O, Fuchs E, Lennartz C, Kido J (2010) High-efficiency blue and white organic light-emitting devices incorporating a blue iridium carbene complex. *Adv. Mater.* **22**, 5003–5007.
44. Wang ZB, Helander MG, Liu ZW, Greiner MT, Qiu J, Lu ZH (2009) Controlling carrier accumulation and exciton formation in organic light emitting diodes. *Appl. Phys. Lett.* **96**, 043303.
45. D'Andrade et al. US Patent 8,557,399 B2.
46. Chang YL, Gong S, Wang X, White R, Yang C, Wang S, Lu ZH (2014) Highly efficient greenish-blue platinum-based phosphorescent organic light-emitting diodes on a high triplet energy platform. *Appl. Phys. Lett.* **104**, 173303.
47. Erickson NC, Holmes RJ (2014) Engineering efficiency roll-off in organic light-emitting devices. *Adv. Funct. Mater.* **24**, 6074–6080.
48. Kim BS, Lee JY (2014) Engineering of mixed host for high external quantum efficiency above 25% in green thermally activated delayed fluorescence device. *Adv. Funct. Mater.* **24**, 3970–3977.

49. Lee CW, Lee JY (2013) High quantum efficiency in solution and vacuum processed blue phosphorescent organic light emitting diodes using a novel benzofuropyridine-based bipolar host material. *Adv. Mater.* **25**, 596–600.

50. Duan L, Qiao J, Sun Y, Qiu Y (2011) Strategies to design bipolar small molecules for OLEDs: donor-acceptor structure and non-donor-acceptor structure. *Adv. Mater.* **23**, 1137–1144.

51. Chang Y-L, Song Y, Wang Z, Helander MG, Qiu J, Chai L, Liu Z, Scholes GD, Lu Z (2013) Highly efficient warm white organic light-emitting diodes by triplet exciton conversion. *Adv. Funct. Mater.* **23**, 705–712.

52. Chang YL, Wang ZB, Helander MG, Qiu J, Puzzo DP, Lu ZH (2012) Enhancing the efficiency of simplified red phosphorescent organic light emitting diodes by exciton harvesting. *Org. Electron.* **13**, 925–931.

53. Staroske W, Pfeiffer M, Leo K, Hoffmann M (2007) Single-step triplet-triplet annihilation: an intrinsic limit for the high brightness efficiency of phosphorescent organic light emitting diodes. *Phys. Rev. Lett.* **98**, 197402.

54. Divayana Y, Sun XW (2007) Observation of excitonic quenching by long-range dipole-dipole interaction in sequentially doped organic phosphorescent host-guest system. *Phys. Rev. Lett.* **99**, 143003.

55. Nakanotani H, Higuchi T, Furukawa T, Masui K, Morimoto K, Numata M, Tanaka H, Sagara Y, Yasuda T, Adachi C (2014) High-efficiency organic light-emitting diodes with fluorescent emitters. *Nat. Commun.* **5**, 4016.

56. Li G, Zhu D, Peng T, Liu Y, Wang Y, Bryce MR (2014) Very high efficiency orange-red light-emitting devices with low roll-off at high luminance based on an ideal host–guest system consisting of two novel phosphorescent iridium complexes with bipolar transport. *Adv. Funct. Mater.* **24**, 7420–7426.

57. Seino Y, Sasabe H, Pu Y-J, Kido J (2014) High-performance blue phosphorescent OLEDs using energy transfer from exciplex. *Adv. Mater.* **26**, 1612–1616.

58. Goushi K, Yoshida K, Sato K, Adachi C (2012) Organic light-emitting diodes employing efficient reverse intersystem crossing for triplet-to-singlet state conversion. *Nat. Photon.* **6**, 253–258.

59. Matsumoto N, Nishiyama M, Adachi C (2008) Exciplex formations between tris(8-hydoxyquinolate)aluminum and hole transport

materials and their photoluminescence and electroluminescence characteristics. *J. Phys. Chem. C* **112**, 7735–7741.

60. Shin H, Lee S, Kim K-H, Moon C-K, Yoo S-J, Lee J-H, Kim J-J (2014) Blue phosphorescent organic light-emitting diodes using an exciplex forming co-host with the external quantum efficiency of theoretical limit. *Adv. Mater.* **26**, 4730–4734.
61. Sun JW, Lee J-H, Moon C-K, Kim K-H, Shin H, Kim J-J (2014) A fluorescent organic light-emitting diode with 30% external quantum efficiency. *Adv. Mater.* **26**, 5684–5688.
62. Park Y-S, Lee S, Kim K-H, Kim S-Y, Lee J-H, Kim J-J (2013) Exciplex-forming co-host for organic light-emitting diodes with ultimate efficiency. *Adv. Funct. Mater.* **23**, 4914–4920.
63. Park Y-S, Kim K-H, Kim J-J (2013) Efficient triplet harvesting by fluorescent molecules through exciplexes for high efficiency organic light-emitting diodes. *Appl. Phys. Lett.* **102**, 153306.
64. Fukagawa H, Shimizu T, Kamada T, Kiribayashi Y, Osada Y, Hasegawa M, Morii K, Yamamoto T (2014) Highly efficient and stable phosphorescent organic light-emitting diodes utilizing reverse intersystem crossing of the host material. *Adv. Opt. Mater.* **2**, 1070–1075.
65. Ng T-W, Lo M-F, Fung M-K, Zhang W-J, Lee C-S (2014) Charge-transfer complexes: charge-transfer complexes and their role in exciplex emission and near-infrared photovoltaics. *Adv. Mater.* **26**, 5226.
66. Blochwitz J, Pfeiffer M, Fritz T, Leo K (1998) Low voltage organic light emitting diodes featuring doped phthalocyanine as hole transport material. *Appl. Phys. Lett.* **73**, 729–731.
67. Walzer K, Maennig B, Pfeiffer M, Leo K (2007) Highly efficient organic devices based on electrically doped transport layers. *Chem. Rev.* **107**, 233–1271.
68. Meerheim R, Lussem B, Leo K (2009) Efficiency and stability of p-i-n type organic light emitting diodes for display and lighting applications. *Proc. IEEE.* **97**, 1606–1626.
69. Pfeiffer M, Beyer A, Fritz T, Leo K (1998) ontrolled doping of phthalocyanine layers by cosublimation with acceptor molecules: a systematic Seebeck and conductivity study. *Appl. Phys. Lett.* **73**, 3202–3204.
70. Pfeiffer M, Forrest SR, Leo K, Thompson ME (2002) Electrophosphorescent p–i–n organic light-emitting devices for very-high-efficiency flat-panel displays. *Adv. Mater.* **14**, 1633–1636.

71. Harada K, Werner AG, Pfeiffer M, Bloom CJ, Elliott CM, Leo K (2005) Organic homojunction diodes with a high built-in potential: interpretation of the current-voltage characteristics by a generalized einstein relation. *Phys. Rev. Lett.* **94**, 036601.

72. Nollau A, Pfeiffer M, Fritz T, Leo K (2000) Controlled n-type doping of a molecular organic semiconductor: naphthalenetetracarboxylic dianhydride (NTCDA) doped with bis(ethylenedithio)-tetrathiafulvalene (BEDT-TTF). *J. Appl. Phys.* **87**, 4340–4343.

73. Mori T, Fujikawa H, Tokito S, Taga Y (1998) Electronic structure of 8-hydroxyquinoline aluminum/LiF/Al interface for organic electroluminescent device studied by ultraviolet photoelectron spectroscopy. *Appl. Phys. Lett.* **73**, 2763–2765.

74. Kido J, Matsumoto T (1998) Bright organic electroluminescent devices having a metal-doped electron-injecting layer. *Appl. Phys. Lett.* **73**, 2866–2868.

75. Schwab T, Schubert S, Hofmann S, Fröbel M, Fuchs C, Thomschke M, Müller-Meskamp L, Leo K, Gather MC (2013) Highly efficient color stable inverted white top-emitting OLEDs with ultra-thin wetting layer top electrodes. *Adv. Opt. Mater.* **1**, 707–713.

76. Chen S, Deng L, Xie J, Peng L, Xie L, Fan Q, Huang W (2010) Recent developments in top-emitting organic light-emitting diodes. *Adv. Mater.* **22**, 5227–5239.

77. Wu C-C, Lin C-L, Hsieh P-Y, Chiang H-H (2004) Methodology for optimizing viewing characteristics of top-emitting organic light-emitting devices. *Appl. Phys. Lett.* **84**, 3966–3968.

78. Smith LH, Wasey JAE, Barnes WL (2004) Light outcoupling efficiency of top-emitting organic light-emitting diodes. *Appl. Phys. Lett.* **84**, 2986–2988.

79. Hung LS, Tang CW, Mason MG, Raychaudhuri P, Madathil J (2001) Application of an ultrathin LiF/Al bilayer in organic surface-emitting diodes. *Appl. Phys. Lett.* **78**, 544–546.

80. Riel H, Karg S, Beierlein T, Ruhstaller B, Rieß W (2003) Phosphorescent top-emitting organic light-emitting devices with improved light outcoupling. *Appl. Phys. Lett.* **82**, 466–468.

81. Huang Q, Walzer K, Pfeiffer M, Lyssenko V, He G, Leo K (2006) Highly efficient top emitting organic light-emitting diodes with organic outcoupling enhancement layers. *Appl. Phys. Lett.* **88**, 113515.

82. Riel H, Karg S, Beierlein T, Rieß W, Neyts K (2003) Tuning the emission characteristics of top-emitting organic light-emitting devices by

means of a dielectric capping layer: an experimental and theoretical study. *J. Appl. Phys.* **94**, 5290–5296.

83. Hofmann S, Thomschke M, Freitag P, Furno M, Lüssem B, Leo K (2010) Top-emitting organic light-emitting diodes: influence of cavity design. *Appl. Phys. Lett.* **97**, 253308.

84. Schubert EF, Hunt NEJ, Micovic M, Malik RJ, Sivco DL, Cho AY, Zydzik GJ (1994) Highly efficient light-emitting diodes with microcavities. *Science* **265**, 943–945.

85. Deppe DG, Lei C, Lin CC, Huffaker DL (1994) Spontaneous emission from planar microstructures. *J. Mod. Optics.* **41**, 325–344.

86. Lee J-H, Kuan-Yu C, Chia-Chiang H, Hung-Chi C, Chih-Hsiang C, Yean-Woei K, Yang CC (2006) Radiation simulations of top-emitting organic light-emitting devices with two- and three-microcavity structures. *J. Disp. Technol.* **2**, 130–137.

87. Canzler TW, Murano S, Pavicic D, Fadhel O, Rothe C, Haldi A, Hofmann M, Huang Q (2011) 66.2: Efficiency enhancement in white PIN OLEDs by simple internal outcoupling methods. *SID Symp. Digest Tech. Papers* **42**, 975–978.

88. Chih-Jen Y, Liu S-H, Hsing-Hung H, Chih-Che L, Cho T-Y, Chung-Chih W (2007) Microcavity top-emitting organic light-emitting devices integrated with microlens arrays: simultaneous enhancement of quantum efficiency, cd/A efficiency, color performances, and image resolution. *Appl. Phys. Lett.* **91**, 253508.

89. Chen C-W, Lin C-L, Chung-Chih W (2004) An effective cathode structure for inverted top-emitting organic light-emitting devices. *Appl. Phys. Lett.* **85**, 2469–2471.

90. Lee H, Park I, Kwak J, Yoon DY, Lee C (2010) Improvement of electron injection in inverted bottom-emission blue phosphorescent organic light emitting diodes using zinc oxide nanoparticles. *Appl. Phys. Lett.* **96**, 153306.

91. Changhun Y, Hyunsu C, Kang H, Lee YM, Park Y, Yoo S (2009) Electron injection via pentacene thin films for efficient inverted organic light-emitting diodes. *Appl. Phys. Lett.* **95**, 053301.

92. Zhou X, Pfeiffer M, Huang JS, Blochwitz-Nimoth J, Qin DS, Werner A, Drechsel J, Maennig B, Leo K (2002) Low-voltage inverted transparent vacuum deposited organic light-emitting diodes using electrical doping. *Appl. Phys. Lett.* **81**, 922–924.

93. Meyer J, Winkler T, Hamwi S, Schmale S, Johannes H-H, Weimann T, Hinze P, Kowalsky W, Riedl T (2008) Transparent inverted organic

light-emitting diodes with a tungsten oxide buffer layer. *Adv. Mater.* **20**, 3839–3843.

94. Thomschke M, Hofmann S, Olthof S, Anderson M, Kleemann H, Schober M, Lüssem B, Leo K (2011) Improvement of voltage and charge balance in inverted top-emitting organic electroluminescent diodes comprising doped transport layers by thermal annealing. *Appl. Phys. Lett.* **98**, 083304.
95. Tsao JY, Crawford MH, Coltrin ME, Fischer AJ, Koleske DD, Subramania GS, Wang GT, Wierer JJ, Karlicek RF (2014) Toward smart and ultra-efficient solid-state lighting. *Adv. Opt. Mater.* **2**, 809–836.
96. D'Andrade BW, Holmes RJ, Forrest SR (2004) Efficient organic electrophosphorescent white-light-emitting device with a triple doped emissive layer. *Adv. Mater.* **16**, 624–628.
97. Eom S-H, Zheng Y, Wrzesniewski E, Lee J, Chopra N, So F, Xue J (2009) White phosphorescent organic light-emitting devices with dual triple-doped emissive layers. *Appl. Phys. Lett.* **94**, 153303.
98. Schwartz G, Pfeiffer M, Reineke S, Walzer K, Leo K (2007) Harvesting triplet excitons from fluorescent blue emitters in white organic light-emitting diodes. *Adv. Mater.* **19**, 672–3676.
99. Sun Y, Giebink NC, Kanno H, Ma B, Thompson ME, Forrest SR (2006) Management of singlet and triplet excitons for efficient white organic light-emitting devices. *Nature* **440**, 908–912.
100. Schwartz G, Reineke S, Rosenow TC, Walzer K, Leo K (2009) Triplet harvesting in hybrid white organic light-emitting diodes. *Adv. Funct. Mater.* **19**, 1319–1333.
101. Jou J-H, Shen S-M, Lin C-R, Wang Y-S, Chou Y-C, Chen S-Z, Jou Y-C (2011) Efficient very-high color rendering index organic light-emitting diode. *Org. Electron.* **12**, 865–868.
102. Kanno H, Holmes RJ, Sun Y, Kena-Cohen S, Forrest SR (2006) White stacked electrophosphorescent organic light-emitting devices employing MoO3 as a charge-generation layer. *Adv. Mater.* **18**, 339–342.
103. Qi X, Slootsky M, Forrest S (2008) Stacked white organic light emitting devices consisting of separate red, green, and blue elements. *Appl. Phys. Lett.* **93**, 193306.
104. Lee T-W, Noh T, Choi B-K, Kim M-S, Shin DW, Kido J (2008) High-efficiency stacked white organic light-emitting diodes. *Appl. Phys. Lett.* **92**, 043301.

105. Lee S, Shin H, Kim J-J (2014) High-efficiency orange and tandem white organic light-emitting diodes using phosphorescent dyes with horizontally oriented emitting dipoles. *Adv. Mater.* **26**, 5864–5868.

106. Reineke S, Lindner F, Schwartz G, Seidler N, Walzer K, Lussem B, Leo K (2009) White organic light-emitting diodes with fluorescent tube efficiency. *Nature* **459**, 234–238.

107. Wang R, Liu D, Ren H, Zhang T, Yin H, Liu G, Li J (2011) Highly efficient orange and white organic light-emitting diodes based on new orange iridium complexes. *Adv. Mater.* **23**, 2823–2827.

108. Chen S, Tan G, Wong W-Y, Kwok H-S (2011) White organic light-emitting diodes with evenly separated red, green, and blue colors for efficiency/color-rendition trade-off optimization. *Adv. Funct. Mater.* **21**, 3785–3793.

109. Meerheim R, Furno M, Hofmann S, Lussem B, Leo K (2010) Quantification of energy loss mechanisms in organic light-emitting diodes. *Appl. Phys. Lett.* **97**, 253305.

110. Sun Y, Forrest SR (2008) Enhanced light out-coupling of organic light-emitting devices using embedded low-index grids. *Nat. Photon.* **2**, 483–487.

111. Hong K, Yu HK, Lee I, Kim K, Kim S, Lee JL (2010) Enhanced light out-coupling of organic light-emitting diodes: spontaneously formed nanofacet-structured MgO as a refractive index modulation layer. *Adv. Mater.* **22**, 4890–4894.

112. Wang ZB, Helander MG, Qiu J, Puzzo DP, Greiner MT, Hudson ZM, Wang S, Liu ZW, Lu ZH (2011) Unlocking the full potential of organic light-emitting diodes on flexible plastic. *Nat. Photon.* **5**, 753–757.

113. Hong K, Lee JL (2011) Review paper: recent developments in light extraction technologies of organic light emitting diodes. *Electron. Mater. Lett.* **7**, 77–91.

114. Koo WH, Jeong SM, Araoka F, Ishikawa K, Nishimura S, Toyooka T, Takezoe H (2010) Light extraction from organic light-emitting diodes enhanced by spontaneously formed buckles. *Nat. Photon.* **4**, 222–226.

115. Ou Q-D, Zhou L, Li Y-Q, Shen S, Chen J-D, Li C, Wang Q-K, Lee S-T, Tang J-X (2014) Extremely efficient white organic light-emitting diodes for general lighting. *Adv. Funct. Mater.* **24**, 7249–7256.

116. Ye S, Rathmell AR, Chen Z, Stewart IE, Wiley BJ (2014) Metal nanowire networks: the next generation of transparent conductors. *Adv. Mater.* **26**, 6670–6687.

117. Li N, Oida S, Tulevski GS, Han S-J, Hannon JB, Sadana DK, Chen T-C (2013) Efficient and bright organic light-emitting diodes on single-layer graphene electrodes. *Nat. Commun.* **4**, 2294.

118. Flammich M, Frischeisen J, Setz DS, Michaelis D, Krummacher BC, Schmidt TD, Brutting W, Danz N (2011) Oriented phosphorescent emitters boost OLED efficiency. *Org. Electron.* **12**, 1663–1668.

119. Reineke S, Walzer K, Leo K (2007) Triplet-exciton quenching in organic phosphorescent light-emitting diodes with ir-based emitters. *Phys. Rev. B* **75**, 125328.

120. Kondakov DY, Sandifer JR, Tang CW, Young RH (2003) Nonradiative recombination centers and electrical aging of organic light-emitting diodes: direct connection between accumulation of trapped charge and luminance loss. *J. Appl. Phys.* **93**, 1108–1119.

121. So F, Kondakov D (2010) Degradation mechanisms in small-molecule and polymer organic light-emitting diodes. *Adv. Mater.* **22**, 3762–3777.

122. Lo MF, Ng TW, Mo HW, Lee CS (2013) Direct threat of a UV-ozone treated indium-tin-oxide substrate to the stabilities of common organic semiconductors. *Adv. Funct. Mater.* **23**, 1718–1723.

123. Wang Q, Sun B, Aziz H (2014) Exciton–polaron-induced aggregation of wide-bandgap materials and its implication on the electroluminescence stability of phosphorescent organic light-emitting devices. *Adv. Funct. Mater.* **24**, 2975–2985.

124. Oh C-H, Shin H-J, Nam W-J, Ahn B-C, Cha S-Y, Yeo S-D (2013) 21.1: Invited paper: technological progress and commercialization of OLED TV. *SID Symp. Digest Tech. Papers* **44**, 239–242.

Index